PRAISE FOR

THE WAVE WATCHER'S COMPANION

"An eye-popping journey . . . liberally peppered with historical and literary references to tickle the palate of discriminating bibliophiles."
—*The Wall Street Journal*

"An entertaining take on a very complex subject . . . by the end, I'd actually learned a great deal." —*GeekDad*

"Pretor-Pinney gives clear and entertaining accounts in his characteristically perky style . . . [he] reminds me of the best kind of science teacher—clever, passionate, indomitable in his determination to share his knowledge." —*The Daily Telegraph* (UK)

"A witty and well-researched volume . . . it's looping digressions taking us on a series of suitably nonlinear journeys along wobbly bridges, earthquake zones, tidal bores, and Neolithic burial sites."
—*The Times Literary Supplement* (UK)

"[Pretor-Pinney] is an entertaining and informal teacher."
—*The Daily Mail* (UK)

PRAISE FOR

THE CLOUDSPOTTER'S GUIDE

"Charming . . . spills over with strange and interesting asides."
—*The Boston Globe*

"Thoroughly readable . . . Never mind the silver lining. It turns out the cloud's the thing.'" —*The Washington Post*

"This book will entice readers of every stripe." —*The Seattle Times*

"Plain old sunshine seems dull in comparison." —*The Economist*

continued . . .

A PERIGEE BOOK
Published by the Penguin Group
Penguin Group (USA) Inc.
375 Hudson Street, New York, New York 10014, USA
Penguin Group (Canada), 90 Eglinton Avenue East, Suite 700, Toronto, Ontario M4P 2Y3,
Canada (a division of Pearson Penguin Canada Inc.)
Penguin Books Ltd., 80 Strand, London WC2R 0RL, England
Penguin Group Ireland, 25 St. Stephen's Green, Dublin 2, Ireland
(a division of Penguin Books Ltd.)
Penguin Group (Australia), 250 Camberwell Road, Camberwell, Victoria 3124, Australia
(a division of Pearson Australia Group Pty. Ltd.)
Penguin Books India Pvt. Ltd., 11 Community Centre, Panchsheel Park,
New Delhi–110 017, India
Penguin Group (NZ), 67 Apollo Drive, Rosedale, Auckland 0632, New Zealand
(a division of Pearson New Zealand Ltd.)
Penguin Books (South Africa) (Pty.) Ltd., 24 Sturdee Avenue, Rosebank, Johannesburg 2196,
South Africa
Penguin Books Ltd., Registered Offices: 80 Strand, London WC2R 0RL, England

While the author has made every effort to provide accurate telephone numbers and Internet
addresses at the time of publication, neither the publisher nor the author assumes any responsibility for errors, or for changes that occur after publication. Further, the publisher does not
have any control over and does not assume any responsibility for author or third-party websites
or their content.

THE WAVE WATCHER'S COMPANION

Copyright © 2010 by Gavin Pretor-Pinney
Chapter illustrations by David Rooney
Diagrams on pages 15, 27, 39, 50, 55, 68, 80, 84, 187, and 190 by Graham White, NB Illustration
www.thewavechannel.org

PRINTING HISTORY
Bloomsbury edition / 2010
Perigee hardcover edition / July 2010
Perigee trade paperback edition / June 2011

Perigee trade paperback ISBN: 978-0-399-53670-0

PRINTED IN THE UNITED STATES OF AMERICA

10   9   8   7   6   5   4   3   2   1

Most Perigee books are available at special quantity discounts for bulk purchases for sales pro-
motions, premiums, fund-raising, or educational use. Special books, or book excerpts, can also
be created to fit specific needs. For details, write: Special Markets, Penguin Group (USA) Inc.,
375 Hudson Street, New York, New York 10014.

THE
WAVE WATCHER'S
COMPANION

Ocean Waves, Stadium Waves, and
All the Rest of Life's Undulations

GAVIN PRETOR-PINNEY

Chapter illustrations by
David Rooney

Additional illustrations by
Graham White

A PERIGEE BOOK

For Flora.

A Note About Measurements

Most of the measurements in this book are in Imperial units (miles, feet, inches, and so on) except for those that are very small, where metric units (such as millimeters and nanometers) have been used instead.

CONTENTS

Wave Watching
for Beginners

One chilly February afternoon my three-year-old daughter, Flora, and I were messing around on the rocks in Cornwall. Normally, this would have been a perfect opportunity for some cloudspotting. But that day was unseasonably clear—in fact, there was not a single cloud to be seen. And as we sat at the edge of the cove, with nothing but the monotonous Atlantic horizon ahead, we found ourselves, by default, watching the motion of the water. At least, I did. Flora just wanted to clamber about on the slippery boulders.

There was nothing dramatic about that day's waves. They weren't barreling breakers throwing up clouds of spray as they slammed into the headland. Nor did they have any of the regularity of the waves you see in your mind's eye, arriving as a steady succession of crests, one after the other, tumbling in regimented fashion up the shore.

I'm late, I'm late, I'm late.

In fact, there wasn't the slightest order to the water's motion. Like rush-hour commuters at a busy station, the little crests passed this way and that, crossing each other's paths chaotically. But, unlike commuters, they passed through and over each other, combining and dividing, appearing and disappearing.

Their movement was mesmerizing. I found myself unable to follow the progress of any individual crest for more than a second. No sooner had I fixated on one than the pesky little peak joined with one coming from a different direction. Then, inevitably, my eye would be distracted by a third wavelet that would sweep through just as the first two vanished.

Talking with Flora, the questions were soon coming thick and fast: "Why are there waves?" "Where do they come from?" "Why do they splash like that?" And although they were childish questions, it was me, not Flora, who was asking them.

Although a clear blue sky had triggered my interest in waves, I now realize that cloudspotting leads naturally to wave watching. You can't stare at the clouds for long before realizing how much their appearance is influenced by waves. I don't mean the waves rolling over the surface of the ocean, but the ones that form up there, within the boundless airstreams of the sky. For the atmosphere is an ocean, too, but an ocean of air rather than of water.

The oceans above and below the horizon are intimately related. As the book of Genesis records, the first thing God did, when getting everything started, was to set the seas in motion:

In the beginning God created the heaven and the earth.
And the earth was without form, and void; and darkness was
upon the face of the deep. And the Spirit of God moved upon
the face of the waters.[1]

The following day, He "divided the waters which were under the firmament from the waters which were above the firmament."[2] In other words, God separated the oceans below from the clouds above with an expanse of air.

This close affinity, if not common lineage, between sky and sea means that a mere cloudspotter is in fact, without even realizing it, a wave watcher, since clouds are often born on waves of air.

These waves take the form of rising and dipping winds, which, though invisible, are revealed by the shapes of the clouds. The "undulatus" species, for example, is either a continuous layer of cloud with an undulating surface or parallel bands of cloud separated by gaps. Such clouds are born in the region of wind shear that occurs between airstreams of different directions or speeds. Undulatus is a beautiful, if common, example of the clouds revealing the waves of the atmosphere.

But the most spectacular example of waves in the sky has to be the rare and fleeting "Kelvin–Helmholtz wave cloud." This snappily named formation looks like a long succession of what surfers call "pipes" or "barrels," but are more accurately described as vortices. It is an extreme example of the undulatus species in which the shearing winds are at just the right speeds to cause the cloud waves to curl

OF INTEREST TO
CLOUDSPOTTERS

OF INTEREST TO
WAVE WATCHERS

THE KELVIN–HELMHOLTZ WAVE CLOUD

Cloudspotters and wave watchers are united by the beauty
of the Kelvin–Helmholtz wave cloud.

over themselves. The fleeting formation appears for no more than
a minute or two before dissipating. While the processes that form it
have little in common with those causing an ocean wave to tumble
upon the shore, this cloud surely sits slap-bang in the middle of the
Venn diagram of cloud and wave enthusiasts' interests.

A wave cloud, or any other cloud for that matter, is a collection of
suspended water particles; but what exactly is an ocean wave? You
may think the answer is obvious: it is a moving mound of water.
But if you do think that, you are not watching carefully enough.
The best way to see that it is not is to observe the effect waves
have on something floating in the water–a sprig of seaweed, for
example.

Before Flora and I abandoned the water's edge, I watched one such tuft of weed rise and dip and duck and weave as it kept pace with the agitated water below. The seaweed seemed less like a rushing commuter and more like a featherweight boxer. As the peaks moved this way and that below it, the bobbing seaweed remained in the same general position. It wasn't swept along with the crests.

When we climbed onto the cliff top, we watched how a boat moved with the passage of the waves. From up there the waves had a wholly different appearance. The chaotic crisscross of crests now looked like just a surface texture, which reflected a glittering path of sparkles below the sun. Beneath these shimmering wavelets, one could see that a far broader, more orderly, pattern of undulations was rolling in toward us from far out in the Atlantic. Each smooth wave was, I'd guess, some 50–60ft from its predecessor, which it followed in a calm, sedate fashion. This caravan of crests could not have been more different from the little peaks that busied across its surface. But, just as they had passed under the seaweed rather than sweeping it along with them, so these gentle giants rolled in under a fishing trawler that was returning with its catch. They didn't drag it toward the land, as they would have done had they been currents of water. Clearly, the water that the boat floated on returned to pretty much where it had started after the wave had passed through.

So if these waves, like the ones in the cove, weren't traveling water, what exactly were they? What was moving from out at sea to the shore?

The answer is energy.

Water is just the means by which energy moves from one place to another. It is the "medium" through which the wave's energy travels. The ocean's surface becomes possessed by energy in the same way that a spiritual medium is supposedly animated by souls from the "other side."

Well, not *exactly* the same.

In fact, not the same at all.

But I rather like the image of the water being like some back-room psychic, wearing jangly earrings and a lot of purple. Placing her gnarled hands on the table, she rises to her feet, animated by the ghost of your dead granny. Her eyes roll, flecks of spittle gather at the corners of her mouth and, she says in a guttural voice that the television's not as good on the other side. Then Granny's spirit moves on, and the old psychic flops back down into her armchair and asks you to cross her palm with money.

Granny's ghost

Does that help explain that an ocean wave is energy passing through water? Perhaps not. In fact, the water doesn't rise straight up and down (like our possessed psychic did) as the energy passes through. Had Flora and I been able to watch the movement of a sprig of seaweed out in the deep water, where the broad, more regular waves were coming in, we might have been able to see the way it moved with the passage of crest and trough. As the wave approached, the seaweed would have been sucked slightly toward it. It would then have risen upward as the crest arrived and, at the highest point, would have moved forward a little with the wave. Then the seaweed would have sunk back down again with the arrival of the trough, pretty much returning to where it had started. With the passage of a wave, the water at the sea surface moves in circles.

It is not easy to picture the water returning to where it started as the energy travels on, so it might be more instructive to describe a wave in a more tangible way, such as its size. There are two dimensions by which to distinguish a ripple from a tsunami: its height and its "wavelength."

A wave's height is the difference in level between crest and trough. Scientists often prefer to use the measurement known as "amplitude." Generally half the wave height, since it is the level of the wave crest compared with that of still water, it makes their equations for modeling waves simpler. But it feels more intuitive to stick with the crest-to-trough dimension of wave height.

The wavelength is a measure of the distance from one crest, or peak, to the next. Although we often think of a wave as a single crest of water (and we use the term to describe any individual peak),

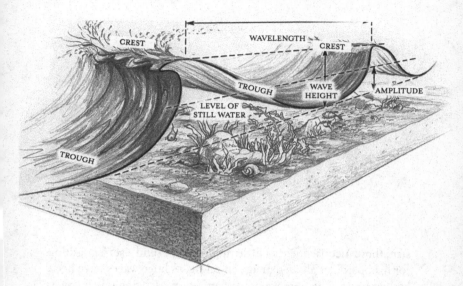

How to size up an ocean wave.

ocean waves never actually exist in isolation. They invariably travel in company, so the term "wave" is often used interchangeably to describe a single crest as well as a succession, or "train," of crests and troughs. Often, as in the Cornish cove, the undulations of the water's surface are so jumbled and confused that it is impossible to discern any clear wavelength. Only when they have the organized appearance of the broad ones we viewed from the cliff-top is it possible to determine whether a wave has a long wavelength, its crests being widely spaced, or a short wavelength, the crests being bunched up.

These two dimensions offer a general measure of the size of waves at any given moment, but say nothing of their movement. And, as any surfer will tell you, waves are all about movement. This is where a wave's "frequency" comes in: that is, the number of crests that pass a fixed point (such as a post protruding from the water) every second. In the case of tiny ripples, such as you might see when you drop a stone into a pond, a handful will scuttle by within a second. But we don't pay much attention to waves of this

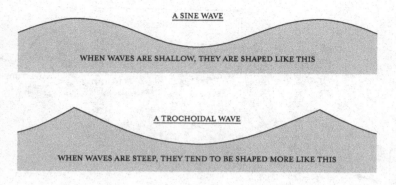

A wave's shape varies with its steepness.

size; these inconsequential little ripples won't send a surfer dashing for his board, let alone damage an oil rig, as huge waves have been known to do. The ones we do pay attention to, being much larger, might have as many as sixteen seconds between the arrival of one crest and the next. In this case, it is rather awkward to talk about the waves having a frequency of one-sixteenth, or 0.0625 crests per second, so the passage of ocean waves is usually described in terms of their "period," which is simply the number of seconds between one crest and the next passing a fixed point.

Besides its size and movement, the other basic characteristic used to describe an ocean wave is its shape. Some rise and fall in broad, symmetrical undulations, with a profile approaching that pure mathematical line the sine wave, as shown in the upper of the two idealized wave shapes above.

But most aren't shaped like that. The steeper a wave, the less like a sine wave it becomes. Rather, it has a "trochoidal" shape. This is slightly less symmetrical than a sine wave, having sharp peaks separated by smooth troughs. But steep doesn't necessarily mean large. The chaotic crests that Flora and I saw in the cove were peaked, rather than rounded. The shape-defining steepness depends upon the height compared with the wavelength, not whether it is a big wave. Even small ones, when bunched up enough, will be steep, and so will have trochoidal shapes.

Clouds and ocean waves have more in common with each other than occasionally looking similar. In fact, breaking waves play a subtle role in the formation of clouds. For as the crests come tumbling in over themselves at the shore, the turbulence causes countless tiny air bubbles to form and burst, releasing a fine mist of water droplets into the air. When the water evaporates, *From waves to* tiny particles of salt are left floating in the air, which can *clouds* become swept up into the atmosphere. These microscopic salt particles are some of the most effective "condensation nuclei" on which most cloud formation depends, acting as seeds onto which the invisible water vapor in the air can start to condense and form tiny droplets that we see as low clouds. I'm not saying that breaking waves cause clouds to form directly above them, just that they ensure that condensation nuclei, important ingredients for cloud formation, are always wafting around the lower atmosphere.

It works the other way, too, for clouds play a role in the formation of waves—at least, storm clouds do. This can seem surprising when you gaze at the waves gently lapping up the beach of some exotic holiday location. They look so calm and tranquil *From clouds to* from the shade of a swaying palm: like the relaxed breathing *waves* of the ocean, each wavy exhalation ceaselessly following the last. But such a graceful arrival belies the waves' troubled upbringing. These serene visitors will often have begun life amidst the chaotic, wind-torn tumult of a storm somewhere out at sea—one that has long since dissipated.

How do waves form in a storm? And how, for that matter, do they change from that choppy confusion into the ordered procession of crests that tumble up the beach toward you? For the answers you need to chart their journey across the ocean, to follow each stage of their development, from birth at sea to foaming death on the shore.

Their biography can be divided into five stages, at each of which the waves have a distinct character.

Let's start at the beginning, with the birth of a wave.

Waves are formed all the time, right across the world's oceans, but it might help to observe their formation in a clutter-free environment: perhaps on a patch of smooth, calm and wave-free sea. In reality, no such place exists. The closest approximations are probably within the "horse latitudes," the bands of the seas that fall between 30° and 35° latitude in both the Northern and Southern Hemispheres, and the "doldrums," the band at the equator, extending 5° or 10° to the north and south. In both regions, the winds can be feeble and inconsistent. Since wind causes waves, there are times when none are generated in parts of these regions. But on any ocean, even during the calmest weather, the glassy surface will still rock with the gentle reverberation of waves that have arrived from storms over other, distant regions.

No such thing as a wave-free sea

The periods of calm can be more persistent in the perpetually high pressure of the horse latitudes. After all, the name is thought by some to derive from eighteenth-century Spanish merchant ships transporting horses to the New World having to jettison their cargo to conserve dwindling water supplies. But we don't want to wait around forever, so let's choose a patch of water in the doldrums. As the poet Samuel Taylor Coleridge famously described them, the winds here can be so feeble and dithering that they leave a sailing vessel "stuck, nor breath nor motion; / As idle as a painted ship / Upon a painted ocean."[3] The word "doldrums" derives from the old English word *dol*, meaning "dull." But these regions experience consistently *low* pressure. This means that their sultry, eerie calm will soon be broken by weather of a very different character that will conveniently turn the gentle, glassy undulations into towering waves.

The warm, humid air around this equatorial belt can lead to intense atmospheric instability. This can cause the air to rise rapidly, its moisture condensing into towering storm clouds. Squalls and storms near the doldrums develop suddenly and can keep growing into enormously destructive tropical cyclones. But we don't need anything as violent as that to bring our waves into being. A simple storm at sea will do the job.

The droplets forming within the building towers of cloud cause the air to warm as they condense, making it expand and rise as it becomes more buoyant. So the air pressure at sea level plummets and surrounding air rushes in to fill the void. This is the wind that gives birth to our ocean waves, and heralds the first stage in their life cycle.

Once the wind's speed has reached a couple of knots, or 3ft per second, the friction it exerts on the water starts to leave subtle imprints. Tiny ripples dance across the surface, each no higher than half an inch or so. Soon, scattered, diamond-shaped ripples, known as "cat's paws," sparkle in the light "where the wind's feet shine along the sea," to borrow a phrase from the Victorian poet Algernon Swinburne.[4] Since a patch of these ripples looks different from the smoother surface away from the wind, it will alert a sailor to an approaching gust before it reaches his sails.

Newborn waves

These incipient ripples are the newborns. They are the very first stage in the life cycle of a wave. The tiny crests come and go with the gusting movements of the wind, but they soon become established as they are nourished by the stiffening air currents of the building storm into an enduring roughening of the water's surface.

And, like all infants, they soon begin to place a strain upon their parents. The roughening of the surface increases the friction between the water and the winds. The air cannot glide over the ocean's surface as easily. Tiny eddies develop just above the capillary waves, resulting in fluctuations in the pressure that the air exerts upon the water. The ripples respond enthusiastically to such stimuli: they lift a crest here and sink a trough there and grow in size.

Their development is always a result of a clash of wills. On the one hand, there's the force of the wind, causing the surface to lift above and sink below its flat, equilibrium level. On the other, there is the water's tendency to resist such disturbances—to return to the calm stability that it enjoys in the absence of the wind. This tendency is, in fact, due to two factors: the water's surface tension and its weight. The tension resists the slight stretching of the surface at a crest and the slight bunching at a trough, while the force of gravity on the water, or weight, pulls down where the surface is lifted above the equilibrium and (by means of the water pressure) pushes up where it is sunk below it. As they try to return the water to a level equilibrium, both forces have the effect of making the wave overshoot, causing a crest to keep sinking and become a trough, causing a trough to keep rising and become a crest. Initially, when our waves are infants, the surface tension of the water is the more dominant force. This natural resistance to the stimulating influence of the wind is what causes the little ripples to progress across the surface.

Once they have risen to an inch in height, our waves are no longer infants. The first stage of their life is complete. No longer can we call them capillary waves. They are now known as "gravity waves" because the weight of the water—the force gravity exerts on it—has become the more significant influence on them. This is now the dominant force in opposing the disturbance of the wind, tending to restore the water to its level and so powering the young waves forward through the ensuing melee.

～

Now into childhood, our waves have entered the second stage of their development, which sees them grow from wavelets to waves, as if from boisterous toddlers to delinquent teenagers. As the wind builds in force and becomes more sustained, so their appearance changes completely from the organized ranks of capillary waves. The crests and troughs now grow agitated and chaotic. They rush this way and that, running into each other, tumbling over each *Toddlers on*   other, like a roomful of toddlers under the dubious guid-
a rampage   ance of a hyperactive babysitter. This confused and irregular ocean surface is known simply as a "wind sea."* The term refers to the sea's rough, choppy surface while it is being whipped up by the wind, as it continues to give more and more of its energy to the water. In the case of a storm, the wind sea is a stage of rapid and sustained growth for the waves.

In fact, the speed of their growth increases as the wind finds larger and larger wave faces to push against. With such a jumble of wave heights and wavelengths coexisting within the same space of water, it is hard to give a representative measure of their

---

* Though, confusingly, it is also sometimes described as just a "sea."

collective size. To describe them in terms of the very highest waves is misleading, since, in the mess of the wind sea, the very highest ones will appear only occasionally. Instead, oceanographers often describe a range, or spectrum, of wave sizes in terms of the "significant wave heights." This is defined as the average of the tallest third of all the waves. It might sound more complicated than the height of the tallest waves, but it is in fact a more useful and representative measure when there is a range.

Before long, the significant wave height has grown to a few feet. They are waves now, rather than wavelets. The fierce storm winds have been anything but a calm, consistent influence upon them. The hyperactive wavelets have grown into aggressive and unruly waves, with steep faces and sharp, trochoidal peaks. They grow angrier and more aggravated under the wind's abusive guardianship, until plumes of foam begin to form at their crests. The third, and most disturbed, stage of their development is about to begin.

～

Now emerging into adulthood, the waves have "badass" written all over them. This third stage is marked by the appearance of foaming lips of white water on the larger specimens. Known as "whitecaps," these are the waves beginning to tumble over themselves, under the relentless, harrying force of the gale.

If storm-force winds blow for long enough, and over a large enough area, they begin to tear plumes of spray, or "spindrift," from the crests. Each wave face becomes marbled with streaks of white foam, "like a wall of green glass topped with snow," as Joseph Conrad described them,[5] though they seem to me *When ocean waves turn nasty* more like the furious spittle of a madman. The waves keep on growing, their significant heights eventually surpassing 16ft.

Now the whitecaps have become commonplace. Mariners sometimes describe them as "white horses," occasionally as "skipper's daughters"—the latter, presumably, because you don't want to mess with them. All this foam indicates that the waves are breaking in deep water. Their crests are being knocked over by the force of the wind. Now, as they continue to grow, the spitting mountains

Foam forms into distinctive streaks down the waves of a storm on the North Pacific, as photographed from the merchant vessel *Noble Star* in the winter of 1989. I'm feeling seasick just looking at this.

of water are most dangerous to ships. Not only are they steepest, they are also liable to break over the ship, bringing tons of seawater crashing down on the deck.

To try to avert such a danger, mariners have, since classical antiquity, had a trick up their sleeves. They have poured fish oil overboard, or hung sacks containing oil-soaked rags into the water, to calm the waves in a storm. It seems that the ancient Greeks considered that this curious effect might be explained by the film of oil that spreads over the surface, reducing friction between wind and water: "Is it, as Aristotle says," wondered the Greek historian Plutarch, "that the wind, slipping over the smoothness so caused, makes no impression and raises no swell?"[6]

Perhaps this phenomenon is what lay behind a wave-calming miracle, described in the eighth century by the English monk and scholar the Venerable Bede. In his *Ecclesiastical History of the English People*, Bede described how a priest setting off on a voyage was given holy oil by a certain Bishop Aidan to chuck in the water if a storm endangered his ship. In this case, the oil was supposed to have had a miraculous effect: it made the wind immediately cease and the storm subside, leaving a balmy, sunny day.[7]

In 1757, the American polymath Benjamin Franklin also became fascinated by the phenomenon, having noticed something peculiar about the waves in the wake of neighboring ships on a transatlantic voyage. The waves behind two of the ships were particularly smooth, compared with the others in the fleet. His captain explained that the cooks must have emptied their greasy water through the scuppers, thus inadvertently calming the waters.

Benjamin Franklin: wave watcher

It obviously stuck in Franklin's mind because, sixteen years later, he described in a letter to a friend, William Brownrigg, an experiment he had performed during a stay in London to study at first hand the effect of oil on wave formation:

> *At length being at Clapham, where there is, on the Common, a large Pond, which I observed to be one Day very rough with the Wind, I fetched out a Cruet of Oil, and dropt a little of it on the Water. I saw it spread itself with surprising Swiftness upon the Surface, but the Effect of smoothing the Waves was not produced; for I had applied it first on the Leeward Side of the Pond where the Waves were largest, and the Wind drove my Oil back upon the Shore. I then went to the Windward Side, where they began to form; and there the Oil tho' not more than a Tea Spoonful produced an instant Calm, over a Space several yards square, which spread amazingly, and extended itself gradually till it reached the Lee Side, making all that Quarter of the Pond, perhaps half an Acre, as smooth as a Looking Glass.[8]*

However, Franklin wasn't able to work out quite why the oil had had this effect. The explanation is a little subtler than the Greek proposal that it makes the water more slippery, preventing the wind from gripping so well.

In fact, the effect the oil has on the water's surface tension is the critical factor. It spreads over the surface of the water as an extremely thin film, or skin, which has a lower surface tension than the surface of water. This drop in surface tension actually makes the water less able to riffle under the influence of the wind and form the less-than-an-inch-high capillary waves.

You'd think that tiny surface ripples would be the least of your worries among the heaving monsters of an ocean storm. But remember that in embryonic form these waves increase the friction between air and water. They give the howling winds a purchase on the faces of the rolling mounds of water, helping them transfer their energy all the more efficiently to the water. By suppressing the surface ripples, the oil can make enough of a difference to the wind's grip to stop an enormous crest from being thrown onto, rather than under, the deck of a ship.

But before you start chucking engine oil in the water the next time you're messing around in a dinghy and it gets a bit choppy, bear in mind that your mini-*Exxon Valdez* will have a minimal effect. Modern petroleum-based oils don't work well. Only organic oils, such as those from the flesh of oily fish, spread far enough, and fast enough, to tame the skipper's daughters.

While we've been distracted on Clapham Common, the storm has continued to howl and, under its abusive guardianship, the waves have grown to significant wave heights, to 40–50ft brutes the size of four-story buildings—with wavelengths of over nearly 800ft. This is now a "fully developed sea," which means that the waves have grown as high as they can under these wind speeds.

The height of the waves whipped up by a storm doesn't just depend on the strength of the wind. Oceanographers have found that there are two other important factors: the area of water over which the wind blows in a consistent direction, known as the "fetch area," and the length of time it does so, or the "fetch duration." These are what will eventually determine whether the storm ever leads to a fully developed sea.

For a good idea of how our waves might appear at the conclusion of their third stage of development, we need look no further than Jan Porcellis's 1620 mini-masterpiece *Dutch Ships in a Gale*, in London's National Maritime Museum.

Porcellis was hailed as the "Raphael of marine painters" by a contemporary, the artist Samuel van Hoogstraten. He helped to popularize seascapes, which had been featured as an artistic subject

"Can we take this back to the boat-rental place now, please?" Jan Porcellis's *Dutch Ships in a Gale* (*c*. 1620) shows the effect on waves of a violent upbringing.

for only about a century, by depicting the heaving surface from close to the water. Characteristically of Porcellis at this time, the painting was small—not much larger than a piece of computer paper—but the low, dramatic perspective must have convinced members of the seventeenth-century Dutch aristocracy viewing it that they were peering through a window onto the marine equivalent of a ferocious tavern brawl. The sheer mayhem of these deranged and uncontrollable waves must have elicited feelings of horror and fascination in equal measure.

Only when the storm eventually passes and the winds die down do our waves enter the fourth stage in their lives. Surprisingly, the calming of the air currents doesn't mean that the furious confusion of peaks and troughs simply settles back down again to a gentle, rocking equilibrium. The waves that were generated in the wind sea continue to travel over the water—but without the need to be pushed along. They've changed from "forced waves," driven by the winds, into "free waves." And how their mood can shift as they mature and enter middle age, finally beginning to distance themselves from their past.

No longer a wind sea, the surface is now what is known as a "swell," which seems an appropriately harmonious name. Although *A comfortable* the storm may have passed, the energy it transferred to *middle age* the water cannot simply disappear. The waves keep going without the need for aerial propulsion. They just roll on, doing their thing. And, as they mature, so the subtleties of their characters begin to emerge.

Waves on the surface of the sea lose remarkably little of their energy to the surroundings once they are up and running. This means that they can travel enormous distances. The little energy that they do lose, a process known as "attenuation," is mostly due to white-capping and, in the case of steeper specimens, air resistance when the wind blows against them. Only the embryonic capillary waves lose much of their energy on account of the viscosity of the water itself. This means that large swells, like those generated by our storm, can travel astonishing distances across the ocean.

This was first demonstrated by Walter Munk at the Scripps Institution of Oceanography near San Diego. Now in his eighties, and still a Professor Emeritus at Scripps, Munk is probably the most respected and renowned oceanographer alive today. During World War II, he had been the first scientist ever to work out a system to predict wave heights. Crucial Allied landings in North Africa, the success of which depended on calm seas, were scheduled according to his forecasts.

In 1957, Munk found evidence that waves reaching Guadalupe Island, off the west coast of Mexico, had originated in storms within the Indian Ocean some 9,000 miles away.[9] A decade later, another study with colleagues from Scripps tracked the progress of ocean swells right across the Pacific from south to north. With highly sensitive wave-measuring equipment positioned at six measuring stations spaced thousands of miles apart, they were able to track the progress of ocean waves. They followed swells originating in storms off Antarctica, recording them as they rolled past New Zealand, Samoa and Hawaii, and over the open expanse of the North Pacific. The same waves finally showed up more than 7,000 miles away in the recording equipment at Yakutat, Alaska, having taken around two weeks to get there.[10, 11]

Long-haul travel, wave style

Over enormous distances like these, the heights of the ocean swells diminished to minuscule levels. The measuring equipment Munk and his colleagues used was able to detect waves a mile in wavelength and an astonishing one-tenth of a millimeter high. But this drop in height is not because the energy ebbs away. It is merely a result of the way waves spread out, fan-like, from their source, the energy imparted to the surface by the storm winds spreading over increasingly greater areas of ocean as they progress.

We all mellow with age. And, compared with the frenetic confusion of the wind sea, the waves in our mature, freely propagating ocean swell exhibit a decidedly relaxed attitude when they cross another swell. As the two enter the same stretch of sea, they simply pass through each other, like friendly ghosts, before continuing on their way without having experienced any lasting interference. The sea surface can look confused as the two swells cross, but they emerge on the other side unaffected by the encounter.

Wise old swells pass through each other and continue with little fuss.

As they move across the open ocean, the jumble of different-sized waves that formed in the wind sea begins to organize itself. This is because of the simple rule that the longer waves travel faster than the short ones. It is rather like a marathon in which the speed of the runners depends on nothing more than how long their legs are—the beanpole runners being faster than the short ones. At the sound of the starting pistol, a huge confusion of runners of different height sets off together. But the rule of the longer the legs, the faster the runner means that they naturally begin to arrange themselves with the beanpoles in the lead and the shorter runners lagging behind.

The same happens with the different-sized waves. As they spread out across the open ocean, the longer ocean waves move faster than the shorter—perhaps 50mph, compared with 30mph—and as a result, the waves in our swell spread out in an orderly fashion.

As their heights decrease—as a result of their energy being spread over wider and wider areas—our maturing waves develop a smoother shape. Gone are the steep, peaked, trochoidal crests of the wind sea. They are now far less precipitous, each crest in the train appearing as a sweeping undulation: "a low broad heaving of the whole ocean," as the Victorian art critic John Ruskin described it, "like the lifting of its bosom by deep-drawn breath after the torture of the storm."[12]

Their smoother appearance now makes them more like the swell in Claude Monet's *The Green Wave*. Monet was something of a pioneer in the deployment of impressionistic techniques for rendering the sea and was described by his fellow Impressionist Edouard Manet as the "Raphael of Water."

If I were Monet, I'd have felt a bit miffed that Manet couldn't come up with a more original compliment.

⌒

The waves actually travel across the open ocean during this fourth stage of their lives in a manner that is far more intriguing and enigmatic than the marathon runners of the last analogy. In fact, the swell is a most peculiar procession of crests. They are in the form

Claude Monet's *The Green Wave* (1866–67).
Never mind the color, just look at that smooth, orderly swell.

of groups of larger waves, separated by gaps, in which the waves are smaller, and sometimes barely even there.

But that's not the weird part. What is so strange is that each individual wave crest travels faster than the group it is in. The crest appears from the calmer water at the back of the group, *A ghost* travels through it and disappears again in the calmer water *marathon* at the front. It is not easy to come up with an analogy to help you picture such odd behavior. The only one I can think of is a train, on which are running not marathon runners, but the *ghosts* of marathon runners.

As this train chugs along in the approach to a station, it happens to be traveling at about jogging speed. Being deceased marathon runners, the ghosts on the train can't actually stop jogging. They therefore appear at the back of each carriage, run through it and disappear again at the front. Anyone waiting in the station for the last ride home will notice the train chugging past at a jogging pace, with the ghostly runners passing through each carriage as it does so. As seen from the platform, the ghosts will be moving twice as fast as the carriages. This, strange to say, is how the waves in the swell move. The crests travel through the group twice as fast as the group itself travels.

The peculiar behavior of our mature waves is the result of the overlapping of waves with similar wavelengths—the longer, faster waves occupying the same water as the slightly shorter, slightly slower waves—so that their crests and troughs add and subtract like this:

IMAGINE TWO OCEAN WAVES
WITH SLIGHTLY DIFFERENT WAVELENGTHS . . .

BOTH MOVING IN
THIS DIRECTION . . . ➞

IF THEY ARE TRAVELING TOGETHER OVER THE SAME REGION OF WATER, THEY WILL OVERLAP . . .

AND, WHERE THEIR CRESTS
AND TROUGHS COINCIDE, THEY
WILL ADD, SO THE SURFACE
WILL HAVE BIG WAVES . . .

BUT, WHERE THEIR CRESTS AND
TROUGHS ARE OUT OF PHASE,
THEY WILL CANCEL OUT, SO THE
SURFACE WILL BE CALMER . . .

WHICH RESULTS IN GROUPS
OF WAVES SEPARATED BY GAPS
OF CALMER WATER . . .

THE GROUPS TRAVEL ALONG
LIKE THE CARRIAGES OF A
TRAIN . . .

AND THE CRESTS TRAVEL
THROUGH THE GROUPS,
LIKE THE GHOSTS OF
MARATHON RUNNERS . . .

THE CRESTS APPEAR IN THE CALMER WATER AT THE BACK OF THE GROUP, TRAVEL
THROUGH IT AND DISAPPEAR AGAIN IN THE CALMER WATER AT THE FRONT –
RATHER LIKE GHOSTS RUNNING THROUGH THE TRAIN CARRIAGES.
ISN'T IT NICE WHEN THINGS ARE SO STRAIGHTFORWARD?

If this peculiar movement of the swell seems rather difficult to understand, perhaps it is best to forget my ghost analogy and just accept that, as the poet Ralph Waldo Emerson put it, "illusion dwells forever in the wave."[13]

~e~

All this talk of peaks and troughs might be considered rather shallow. What about the water below the surface when a wave passes through?

Do you remember how, with the passage of the wave, the water at the surface moves in a roughly circular path—how it returns to near its starting point as the wave moves on? In fact, the water below the surface moves in the same way, except that the circular paths become smaller the deeper they are. At a depth equal to half the wavelength, these circular orbits diminish to nothing. Below this "wave base" the movement of the water is negligible as the wave passes overhead. This is why submarines can avoid the effects of even the fiercest storms merely by diving to 500ft or so.

In fact, waves do occur deeper below the surface, where they are often far larger than the ones on the surface. Dubbed "internal waves," these are the lumbering giants of the ocean. They can occur in the murky depths, wherever an abrupt boundary between layers of differing water develops. If layers of water have quite different densities, perhaps because one is much warmer than the other, or it is much saltier, the boundary between the two behaves rather like an underwater equivalent of the ocean surface. Waves can roll along this boundary, hidden from view on the surface.

These internal waves are set in motion by the action of the

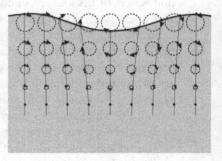

DIRECTION OF WAVE MOVEMENT

The water moves in circular paths as the wave passes, the circles becoming smaller and smaller with depth.

tides, rather than that of the wind. And they are often larger than their surface equivalents—a lot larger. It is not uncommon for internal waves to form with wavelengths of 12 miles and heights in excess of 650ft.

Submarines may be able to dive deep enough to avoid storm waves at the surface, but they cannot escape the influence of the *Monsters* ones lurking below. In the 1960s, one such internal wave *of the deep* smacked into a Russian submarine as it was trying to pass secretly through the Strait of Gibraltar, sending it crashing into an oil platform. This must have caused some red faces among the crew.

～℮

Waves are the expressions of the ocean's moods. Placid, tranquil and benevolent is the sea that caresses the shore and rocks your dinghy in a cradle of gentle undulations. But for a sense of Nature at its most terrifying, nothing comes close to a violent storm at sea. The expressive character of waves has always made the ocean an abundant fishing ground for those in search of a good metaphor.

The seafaring adventures of Odysseus, as described by Homer, are awash with his struggles against the various tempests that Poseidon, the god of the sea, unleashes upon him. *The Odyssey* helped to establish the enduring motif of man as mariner, crossing the storm-tossed seas on the "journey of life" in search of the tranquil waters of his journey's end. But for the classical playwrights, and poets generally, stormy waves have the upper hand. Man's confrontation with the ocean was invariably an against-the-odds battle fighting the whim and fancy of the gods—one in which his heroism and bravery could be put to the ultimate test. "Wonders are many, and none more wonderful than man," wrote the Greek playwright Sophocles 250 years after Homer, "the power that crosses the white sea, driven by the stormy south wind, making a path under surges that threaten to engulf him."[14]

The ceaseless rise and fall of waves seems to echo the arcs and cycles of life. Is this why wave watching is so good at putting life into perspective? As you might expect, a sixty-six-year-old Walt Whitman, contemplating his navel, while gazing out on the waves rolling in at Navesink, New Jersey, was not too bothered about whether they were spilling or collapsing breakers:

> *By that long scan of waves, myself call'd back, resumed upon myself,*
> *In every crest some undulating light or shade—some retrospect,*
> *Joys, travels, studies, silent panoramas—scenes ephemeral,*
> *The long past war, the battles, hospital sights, the wounded and the*
> *   dead,*
> *Myself through every by-gone phase—my idle youth—old age at hand,*
> *My three-score years of life summ'd up, and more, and past,*
> *By any grand ideal tried, intentionless, the whole a nothing,*
> *And haply yet some drop within God's scheme's ensemble—some*
> *   wave, or part of wave,*
> *Like one of yours, ye multitudinous ocean.*[15]

And then, of course, there's the appeal of the wave simply as a shape. Many artists have claimed that the smooth, sinuous line of a wave is one of the most beautiful forms, mirroring, as it does, the recumbent female figure. The English painter William Hogarth included a serpentine curve in a self-portrait that he painted in 1745. The line, in the shape of a wave, appeared carved onto an artist's palette in the bottom left-hand corner of his portrait, with the words "The Line of Beauty and Grace" below it. When the painting became known, Hogarth was inundated with requests for an explanation of what seemed like a cryptic clue. In an effort to answer, the artist then wrote a treatise on aesthetics called *The Analysis of Beauty*. "The serpentine line," he explained, "by its waving and winding at the same time different ways, leads the eye in a pleasing manner along the continuity of its variety."[16] Observing such lines as "formed by the pleasing movement of a ship on the waves," Hogarth added, gave the eye the same pleasure as we derive from following "winding walks and serpentine rivers."[17]

Had roller coasters existed, Hogarth might have mentioned these too. Don't we love them because of the wave-like way the track curves? Our lives are full of ups and downs, even if in real life the highs and lows are far less compressed. We are beset by "uphill" challenges that take us to new places, new heights—to the

A cryptic clue from a closet wave watcher in William Hogarth's
self-portrait, *The Painter and His Pug* (1745).

crest of the wave—before the inevitable, sickening, white-knuckle descent.

Isn't it this aspect that has made the phrase "emotional roller coaster" such an enduring cliché? I do admit, though, that when we plunge headlong into a real-life emotional trough, we generally don't stick our hands in the air and squeal like banshees.

Could you not say that the human body is like a wave? Apparently, once you reach old age, your body can contain none of the molecules it did when you were a newborn. As you *Stretching wave* grow by incorporating what you consume, every ingre- *metaphors to the* dient of your infant body can eventually be replaced: all *breaking point* the particular oxygen, carbon, hydrogen and nitrogen atoms, and the other elements that were your nascent body, will have been replaced. You might say that we borrow the air, water and food we consume in the very way that an ocean wave borrows the water it passes through.

In this sense, you and the wave seem rather alike. Were you able to freeze-frame an ocean wave as it was rolling to shore, it would be tempting to say that the mound of water before you, held in magical suspension, *is* the wave. But waves aren't frozen in time, and in reality the water within the wave at any moment will be left behind an instant later as the wave rolls on. Though the timescales are very different, a wave passes through the medium of the water in much the same way that you pass through the "medium" of all the physical bits of your body.

Clearly, the contemplation of the ocean's undulations can have a disconcerting effect on you. It can set off decidedly hippyish thoughts. Without realizing it, you can be swept out into a Zen-like spiritual reverie on how everything is, like, you know, intercon-nected, deep down, man.

⌒

And so we come to the fifth and final stage in the life of our waves. Perhaps they have traveled in the peculiar grouped arrangement of wave adulthood for many hundreds of miles. Only as they approach landfall do they make one more transformation. It can

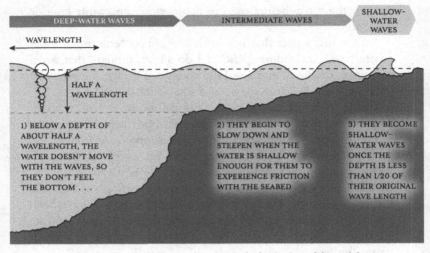

At a depth of half the wavelength, it's the beginning of the end for
our aging waves.

be the most dramatic of all, not least because it marks their death.
This is the stage with which landlubbers are most familiar, when
the waves release their energy in a churning, foaming crash onto
the shore.

Their swan song begins as they enter shallower water. Where
the wave base—that half-wavelength depth at which the water is
only just moving with the wave—first makes contact with the rising
seabed, the waves "feel" the ground beneath them. The progress of
their bases is slowed by friction with the seabed. As they slow, they
bunch up and steepen, so that their shapes change from smooth
undulations back into the sharp, trochoidal peaks of their youth,
when they were harassed by the fierce storm winds of their troubled adolescence.

The transformation from deep-water to shallow-water waves
is complete when the depth has diminished so much (to around
one-twentieth of the wavelength) that the water is no longer able
to move in circles. The motion below the surface is restricted to
increasingly flattened ovals, as there is less and less water for the
wave to travel through. The orbits are squeezed more and more

until the water below the surface is moving almost exclusively forward and backward.

Now one rule comes to dominate their behavior: the shallower the water, the slower they travel. This simple law governs the glorious and dramatic displays as they break into a cascade of foam.

It works like this. Due to the gradient of the seabed, the crests at the front of the wave train slow down before those behind. And, just like a marathon runner stumbling so that those behind fall and ride up on top of each other, so the undulations in the water are collapsed. As the waves are squeezed, there is nowhere for the water to go but up.

If the gradient is just right, and the waves have enough energy, they can rise up so dramatically that they become unstable: below the water, the wave's feet slow down, while the top keeps going, and the wave trips over itself, causing the crest to pitch forward and crash over on itself.

Oceanographers tend to divide breaking waves into three types: "spilling breakers," "plunging breakers" and "surging breakers." Which way a wave breaks depends on the gradient of the *Getting to know* seabed. When the slope of the beach is very shallow, the *your breakers* waves crumble at the crests as spillers. These fringes of white water stretch down from the lip along the front of the wave, making it look as if it is wearing one of those Tudor ruffs.

The waves depicted in *Sennen Cove, Cornwall*, painted by John Everett in 1919 (see next page), are spilling breakers. Everett made extensive voyages around the world, often aboard merchant vessels as a working member of the crew, in order to study and paint waves. He hasn't had the acknowledgment he deserves, and I feel compelled to call him the "Raphael of Spilling Breakers."

Plunging breakers form when the slope of the beach or reef is steeper, and they are the most beautiful of the three types. The lip of the wave is thrown forward so that it curls over to form a tube before crashing to the water below. At their most impressive, these breakers are the "barrels" that surfers ride within, the canopy of water thrown over their heads as they disappear from view.

Surging breakers, which occur on the steepest gradients of the seabed, look completely different. They are hardly breakers at all.

Spilling breakers wear ruffs of white water, as in John Everett's
Sennen Cove, Cornwall (1919).

The water just sloshes up against the steep shore and back again,
like water at the end of the bathtub as you sit yourself down with
a thump. Without any white-water ruff or cascading canopy, these
are the most naked of breakers.

Some textbooks talk about "collapsing breakers," which are
halfway between plunging and surging, but that's getting rather
too nit-picky. In fact, there is a continuum of breaking styles, with
no distinct boundaries between them. Whether we divide them up
into three or four—or, indeed, ten—categories, they are no less arbitrary. A single crest can also break in more ways than one as it rolls
up to the shoreline—spilling over here, before becoming smooth
again and plunging over there, before a nudist streak of a surge at
the end. It all depends on the changing depth near the shore due
to the underwater topography, the rising and falling of the seabed,
known as the "bathymetry." The urge to divide the waves into types
like this reflects our pervading desire to dissect and categorize the
world, to make the continuous digestible. (The same could be said
of my dogged attempt to neatly divide the wave's life into five
quite distinct, and completely separate, stages.)

Whatever the particular style of their demise, our waves finally die on the unyielding shore, as their energy dissipates. They are gone in a tumble of white water; lost, to quote that famous Matthew Arnold poem "Dover Beach," within "the grating roar of pebbles which the waves draw back, and fling, at their return, up the high strand . . . and bring the eternal note of sadness in."[18]

Journey's end

And so the biography of our waves is complete.

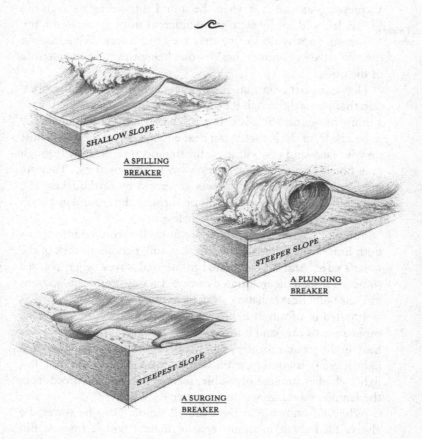

SHALLOW SLOPE

A SPILLING
BREAKER

STEEPER SLOPE

A PLUNGING
BREAKER

STEEPEST SLOPE

A SURGING
BREAKER

And you thought they were all just waves?
Wave watchers should learn the subtleties of wave-breaking.

But perhaps it is a little early to begin mourning.

Energy never expires—all it can do is to change from one form to another. When the waves come crashing onto the shore their energy doesn't just disappear. It keeps on traveling, but in different forms. That "grating roar of pebbles," for instance, is part of the waves' energy transformed into sound.

And sound is a type of wave.

It is a wave not of rising and falling water but of differences in air pressure—at least it is when the sound is traveling through the *A wave is* air. It could hardly seem more different from the waves on the *reborn* ocean, so why do we say they are both waves? What else do the two have in common besides one being the energetic afterlife of the other?

How else do the crashing ocean waves live on? There is the vibration that you feel through the ground when the sea is rough. Lie for a moment against the glistening, black cliff face—far enough away to be safe from the breakers, but near enough to feel their salty mist on your face—and you will sense the tremors reverberating through your body. These vibrations are known as "microseisms." They are mild versions of the shock waves generated by earthquakes. The energy of the breaking waves rolls on through the ground in a form that is subtler, but is a wave nonetheless.

Some of the energy from the ocean swell also turns into heat—both heat in the water and heat in the sand, pebbles or rock of the water's edge. And heat is related to infrared waves: when you see someone filmed on an infrared camera, they are visible on account of their body heat radiating away from them.

Infrared is a form of light—one that we can't see, though some other animals can—and it too is a form of wave. If the ocean waves heat up the ground slightly as they crash against it, then the infrared light it emits is another, even subtler, form of wave afterlife. But light, whether infrared or visible, seems even *more* divorced from the familiar waves we see on the water.

While I'd known that these are all supposed to be waves, I'd always filed them in quite separate mental compartments. But there, at the seashore, the pigeonholes had disintegrated. Amid the glorious death of our ocean waves, the energy rises phoenix-like to live on as other forms of wave. The foaming breakers that tumble onto the beach mark not the end of our wave biography, simply the conclusion of its first chapter.

Waves out at sea were all very well but I had realized that the shore was where it was happening. The waves that Flora and I had watched in Cornwall had whetted my appetite. I now realized what would help me understand waves, and the seemingly mysterious role they play in the world. I would need to undertake an in-depth investigation of waves crashing ashore. I would need to immerse myself in them. I would have to take a holiday to that wave watcher's mecca: Hawaii.

A wave watcher is born

Sorry, did I say *holiday*?

What I meant to say was "research trip."

The First Wave

The only problem was my timing. The Hawaiian waves are at their most impressive when winter storms track across the North Pacific, sending enormous swells on a collision course with the island chain. The perfect time to observe such magnificence was in December and January, I discovered. Which was a bit of a problem, given that it was already late February.

So I decided I'd start wave watching a little closer to home. I soon realized that I need look no farther than the mirror, because anyone who thinks that waves only happen "out there" is quite wrong. In fact, they are constantly traveling through our bodies. Like most animals, we humans rather depend on them.

Waves are quite literally at the heart of human existence. They are the very means by which blood courses around our bodies. For your heart to pump the 4,300 gallons it does in any twenty-four-hour period—cycling oxygen-rich blood through your arteries, veins, and organs—it has to beat 100,000 times. Each and every one of those beats takes the form of a wave.

Muscular contractions of the heart seem so different from undulations traveling over the surface of the water that you might wonder how it can even make sense to describe both as waves. What can the beat of your heart possibly share with the ripples that spread across the surface of your bath as the soap slips from your hand and drops into the water?

Both are forms of traveling oscillations, or vibrations. As one region seesaws between different states, it causes the region next door to start, so that the pattern of movement spreads. In your bath, the falling bar of soap disturbs the surface of the water, causing it to oscillate between dipped and raised levels, and this disturbance spreads in expanding rings. In your heart, the spreading oscillations are of the muscle cells contracting and expanding. Like the changing surface of the water, these contractions spread from one region of the heart tissue to another, though they do so in a very different way.

Tiny electric currents drive the waves of movement that form the beating of your heart. Each cell within the muscle tissue contracts when stimulated by an electric pulse. But in order for the heart to pump your blood efficiently, these contractions need to pass rapidly down through the walls of the heart in a coordinated fashion. The current itself is initiated by a clump of "pacemaker cells" at the top

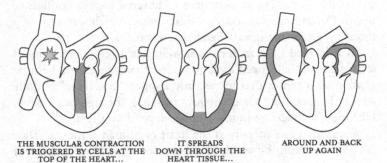

THE MUSCULAR CONTRACTION IS TRIGGERED BY CELLS AT THE TOP OF THE HEART...   IT SPREADS DOWN THROUGH THE HEART TISSUE...   AROUND AND BACK UP AGAIN

Every beat of your heart is an exquisitely coordinated muscular wave.

of your heart, which produce a small electric "shock." This electrical activity spreads down through the muscle, with each cell contracting and passing on the electrical current to its neighbors.

After each cell has fired, it becomes momentarily unable to do so again—as if it is exhausted, and having a rest. Known as the "refractory period," this delay in the cell excitability, which lasts between one-tenth and one-fifth of a second, elegantly ensures that the wave can spread only once through the muscle tissue. Until, that is, the pacemaker cells spontaneously fire up again, starting the wave of your heartbeat once more.

The sterling work done each day by your "household divinity"—as the seventeenth-century physician William Harvey described the heart—is equivalent to the effort required to lift a 2lb weight about twice the height of Mount Everest.[1] (And without the need for a team of Sherpas.) To perform such a feat, timing is crucial. *A cardiovascular* For the heart's four chambers each to fill with blood and *triumph* pump it the right way around the system, they have to contract and expand in a very synchronized, coordinated manner. The two on the right of your heart pump blood through your lungs to oxygenate it. The chambers on the left of your heart then pump the oxygenated blood around the rest of your body. And this timing depends critically on the electrical signals spreading through the muscle tissue being the right shape of wave: one that starts at the closed end of the chamber and progresses evenly across the muscular tissue toward the valve that blood needs to be pumped through.

Hearts don't always work perfectly, however, for if the wave patterns develop abnormal forms, your heart's ability to pump becomes compromised. A target-shaped wave, like the one from the "plop" of your soap into the bath, is just the shape of wave you want to avoid. The same must be said for a spiral-shaped wave, which, in liquid form, is nothing more fancy than the miniature "high tide" that travels around the wall of your teacup as you stir in your lump of sugar. Target- or spiral-shaped waves developing within your heart muscle ruin the careful timing upon which the whole system depends. They lead to a condition generally known as "arrhythmia." Though not as common a cause of heart attack as the furring of the arteries supplying the heart cells with oxygen and nutrients, the effects of arrhythmia range from an occasional and mildly uncomfortable heart flutter, which is no reason for concern, to serious and repeated misfiring, leading to major heart attacks and sudden death. And then there are situations in which the pacemaker cells fail to start the electrical signals properly. This condition can be treated with an artificial pacemaker, which delivers a steady rhythm of little shocks to set the waves off at the right moments.

The target- and spiral-shaped waves that develop within unhealthy heart tissue result from the electrical signals, and the subsequent contractions, failing to spread evenly and efficiently through the muscle tissue. They can be caused by various problems. Sometimes a region of normal muscle cells will develop the cellular equivalent of an identity crisis, and act as if they were pacemaker cells, starting their own mistimed waves. Or the spread of electrical signals can be hindered or slowed by tissue damage or blood clots, rather like a pier or harbor wall disrupting the pattern of ocean waves.

Either way, the result is a condition known as "re-entry dysrhythmia." At its most serious, this leads to a life-threatening emergency in which multiple and uncoordinated waves skitter round and round the cardiac tissue like a sort of muscular feedback, which has the disastrous effect of making the heart tremble rather than contract. (This is when doctors in medical dramas shout things like, "OK, he's gone into v-fib. We have an emergency. Where the hell did you put that defibrillator?") A shock of direct current needs to be administered to the heart within seconds to clear the chaotic electrical signals. A bit like rapidly rebooting a computer, this, hopefully, starts the heart pumping efficiently again.

Yet heartbeats are just one of the many muscular waves that are constantly passing through your body. These may not be the types of wave to get well-toned surfers excited, but they should, for they are the waves on which our very lives depend. Which is, presumably, why these muscular contractions are involuntary— and therefore why most of us don't even realize we are producing them.

The "peristaltic wave," for example, takes the food you swallow down through your esophagus and from there into your stomach.

That same wave of muscular contraction then carries the food from your stomach and into your small intestine to be digested.

These waves are your body's internal transportation. Some are subtler than others. Take the rippling movement of the tiny hairs, or cilia, that line your windpipe. These perform the most civilized of your internal muscular waves, powering a clever process known as the "mucociliary escalator." This sounds a great *Slimy escalators* deal more elegant than the reality: the inside of your windpipe is coated in mucus that captures any particles of dust and pollution that you may breathe in. This sticky layer does a great job of catching these aerosols before they damage your lungs. But how do you then get the mucus and its cargo back up and out of your windpipe without the need to hawk at full volume like some ill-mannered oaf?

By means of ciliary waves, of course. The tiny hairs constantly oscillate just out of step with their neighbors to produce waves reminiscent of those that you can watch sweeping down the legs of millipedes when they are scuttling along in a hurry. The coordinated microscopic movement carries the mucus and its cargo up the windpipe to the larynx. The ciliary waves have done their job. Whether the result is then politely swallowed or crudely coughed up has nothing to do with waves. That is purely the result of what your parents have taught you.

All these internal muscular waves are far too important to be left to conscious control. Just imagine if you had to *remember* to coordinate your peristaltic waves and your mucociliary escalator, while also ensuring that you haven't sent any spiral waves around the old ticker? It would be like playing the most stressful arcade game in the world. With all that on your mind, you would be a hopeless dinner-party guest. You'd spend the whole evening in silence with a look of strained concentration on your face, and would almost certainly never be invited again.

⌒

Any wave watcher worth his salt should know the difference between the three basic forms of wave. These differ in the direction of movement of the medium—the water, for example, through which an ocean wave travels, or the air that transmits a sound. They are "transverse," "longitudinal," and "torsional" waves.

Although the movements are fascinating, their names, I'm sorry to say, risk setting off a wave of boredom. In an effort to avoid this, I'm going to try to demonstrate how they differ from each other by means of animals that use these forms of wave to move around.

I'm not talking about the animals that enjoy surfing ocean waves, although dolphins and porpoises are certainly the world's most accomplished surfers.

Nor am I talking about this:

Buddy, a Jack Russell terrier, competes in the small-dog heat of the 3rd Annual Surf Dog Surf-a-Thon, held in Del Mar, San Diego, California.

No, I mean creatures that contort their bodies into muscular waves as a means of self-propulsion. The purest examples of these body waves are found in the movement of snakes. Since they are extremely successful predators, it can't be such a bad way of getting around. And, as it happens, snakes provide a particularly good demonstration of our first type—transverse waves.

Transverse waves are snake waves because they are ones made up of oscillations perpendicular (at right angles) to the direction the wave is traveling. The shape of the wave moves forward on account of the medium going up and down or from side to side. When a snake moves in this way it is known as "serpentine" locomotion. Although it's actually one of several means of snake propulsion, serpentine locomotion is the basic mode, the one that any self-respecting species needs to master, be it a king cobra or a grass snake. With the full length of its body in contact *Transverse, or* with the ground at all times, the reptile forms an S-shape, *snake waves* undulating its body from side to side to form a wave shape. It makes this muscular wave travel from tip to tail down the length of its body, while pushing sideways against the rough ground to propel itself forward. The undulations trace a continuous, sinuous line over the ground, every point along the snake's body following the same path as the one in front since the body wave travels down its length at the same speed it slithers forward.

A snake will use this mode when it is traveling on rough ground, with plenty of solid things, such as twigs, stones, and other irregularities, on which to get a foothold . . . or, rather, a wavehold. But most species of snake can shift between different styles of movement, rather like horses changing from trotting to cantering to galloping. The mode of movement the snakes use depends on the surface they are traveling over and how fast they want to go.

When the ground is particularly loose, such as shifting sand or slick mudflats, many species use transverse waves in a far flashier way, by switching to a sort of 4x4 version of serpentine locomotion, known as "sidewinding." This is the way to get around where the fluidity or smoothness of the surface renders serpentine locomotion a waste of slithering effort because there is minimal resistance against which to push. As you would expect, desert snakes are particularly adept at this technique, notably the member of the pit viper family that is actually called a sidewinder.

The movement is like a 3D twist on side-to-side serpentine locomotion. For while the snake produces this lateral undulation, its body simultaneously ripples up and down. By combining vertical and horizontal waves at the same time, it moves in a graceful,

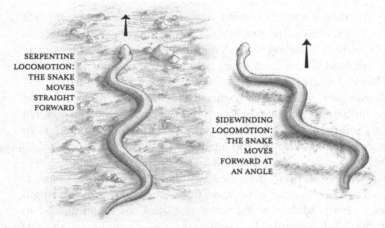

SERPENTINE LOCOMOTION: THE SNAKE MOVES STRAIGHT FORWARD

SIDEWINDING LOCOMOTION: THE SNAKE MOVES FORWARD AT AN ANGLE

Snakes are masters at the use of transverse body waves to get around, and none more than the supreme masters that are able to "sidewind."

corkscrew-like motion with only two or three sections in contact with the sand at any moment, which is helpful when the ground is very hot. A sidewinding snake moves at an oblique angle to the direction it is pointing and leaves a distinctive track in the sand: a succession of J-shapes. It is the most mesmerizing of all the snake movements, as well as the most ostentatious use of transverse waves by any animal.

Many species of snake can also swim, although of course there is even less for them to push off against in the water. Here, they resort to another twist on side-to-side serpentine locomotion, which is known as "anguilliform swimming." This aquatic version differs from the terrestrial body wave in a few crucial aspects. For instance, the undulations become broader down the length of the snake. In other words, the amplitude of the waves increases as they pass down its body so that the tail wiggles from side to side far more than the head. Also, with nothing solid to push against in the water, the waves travel down the snake's body much more rapidly than it glides forward through the water.

Sea snakes, which inhabit warm waters such as the coastal regions of the Indian and Pacific oceans, are masters at this form of movement. If you see one, it is probably best to head in the other direction, by whatever means of propulsion is at your disposal, for they are some of the most venomous species of all. They evolved from terrestrial ancestors but have adapted to a life on the high seas. All have flattened tails, and some—such as the yellow-bellied sea snake—have developed bodies that are slightly taller than wide. Both features help to give them more propulsion through the water. Sea snakes are the only type that has been observed performing the rather flashy Michael Jackson–like swimming maneuver of reversing through the water by making the body waves travel from tail to head. Whichever way the undulations travel, they take the form of transverse waves.

It's not just snakes that find this a useful way of swimming. Eels, lampreys, and hagfish employ the same sort of waves down the length of their bodies to propel them through the water. Rays send them down the sides of their wings: in some cases, as graceful, broad sweeps; in others, as ranks of skittish ripples.

That rays' undulations go up and down, rather than from side to side, makes them no less transverse waves. While fish generally use waves of side-to-side muscular contraction in their tails to kick themselves forward through the water, aquatic mammals, such as whales, dolphins, and seals, generally kick their tails up and down. Mermaids would presumably do the same.

Whether side to side or up and down, they're all transverse waves, for the undulations are perpendicular to the line that the waves travel down the body.

~e~

If waves course through the human body beyond conscious control, what about the conscious control itself? Do they make an appearance on both sides of the Cartesian divide, to play a part in your thinking, too?

Your brain certainly does make use of waves. These are not waves of muscular contraction but of tiny, split-second electrochemical reactions: neurons firing.

It is well known that the electrical impulses traveling along your nerves convey information to your brain from the senses around your body, and also that the traffic is two-way, since your brain sends signals down the nerves to control your muscles and glands. These nerves consist of bundles of special cells, called neurons, each of which has a thin tube, or "axon," connecting a cell body *Brain waves* at one end and a structure of branches at the other. Each of these links to a further cell, be it another neuron or a different type, by means of a tiny junction, or synapse. Though often less than a millimeter long, the axons can occasionally be much longer, as in the case of your sciatic nerve, which extends the length of your leg. The way the signal passes down the length of the neurons is as an electrochemical wave.

You could think of the pulse traveling down the neuron like the wave from a splash traveling down the channel of a narrow stream. Both the walls of the axon and the banks of the stream act as "wave guides" that direct the pulses along a fixed route. Having said that, the two types of wave could hardly be more different from each other. Rather than physical undulations, these pulses, so fundamental to the communication of signals around your body, vary by way of shifting voltages, produced by the chemical reactions within the neurons.

It is not just the neurons within the nerves running to and from your brain that operate by means of these electrical waves. So do those within it. As the hub of your central nervous system, your brain is an intricate network of neurons, each of which is a wave guide, channeling electrochemical pulses from one end to the other.

But there are also far subtler forms of brain waves. These don't travel within the individual neurons, but manifest as swathes of activity that sweep *across* broad regions of the brain, a little like those waves you see spread over wheat fields in the wind. These waves don't consist of neurons firing, but of them getting *ready* to fire.

When a neuron becomes "depolarized," its likelihood of firing increases. This is a little like a person becoming excited, and thereby more likely to start shouting. A growing body of evidence suggests that one way in which the brains of mammals work depends on "waves of excitation" sweeping across the brain.[2] As these waves pass across a region of the brain, the neurons there become more likely to fire. You could think of them like waves of excitement sweeping through a crowd just before a band comes on stage; the individuals each become more likely to shout, to wave their hands in the air, or to dance.

But why might the neurons in animals' brains produce these depolarization waves? And what do the waves of excited neurons actually look like?

~

Amazingly, neuroscientists are now able to watch the changing colors of waves as they sweep over tiny, exposed regions in the brain of an anaesthetized animal. This is because the waves are rendered visible by a special dye. It bonds to the neurons and changes color depending on their "field potentials," which is an electrical measure of how likely they are to fire. The dye renders visible the waves of excitation that sweep across the surface of the animal's functioning brain. They travel very fast and are recorded by a digital camera that registers the shifting hues over an ⅛ in-wide exposed patch of the brain. While the dye has been in use for more than thirty years, only recently has the photographic equipment become sensitive enough to study accurately how the waves move. It turns out that they sweep over the brain tissue in distinctly familiar patterns.

"From what we have observed, there seem to be two basic types of wave shape," explained Professor Jian-young Wu, of Georgetown University's Medical Center in Washington, DC. He has used the technique to study waves on the brains of rats. "One shape is a target wave and the other is a rotating or spiral wave."

Wait a minute! Weren't those the wave shapes that led to cardiac arrest when they spread across the heart muscle? It seems that, in the case of the brain, these wave shapes on a small scale are not a cause for concern. In fact, Wu believes that they are a basic pattern of mammal brain activity. The waves have been observed across many regions of the neocortex, which is the outer layer of animals' brains. This region is involved in higher brain functions, such as processing information from the senses, making the body move, engaging in conscious thought and, in the case of humans, using language.

"Waves have been observed during almost every type of cortical processing examined by voltage-sensitive dye imaging," Wu told me. These waves can flit across the outer neocortices of animals as diverse as turtles, guinea pigs, salamanders, and monkeys. They appear in the conscious animals whether they are stimulated with smells, sounds, lights, or whisker twitching.

Wu has also found that spiral waves race across a rat's brain while it is nodding off. "We can guess that these little spiral waves are generated by very local neuron interactions, and might be a way to help the cortex to escape from the control of the thalamus." (The thalamus is that part of the brain below the neocortex associated with regulating consciousness and alertness. In other words, Wu speculates, these brain waves might rotate over the neocortex like tiny neuronal windshield wipers, detaching reasoning faculties from the stimulation of the thalamus, and thereby allowing the poor animal to drift off to sleep.) He added: "We feel that waves like these are candidates for a means by which complex mental processing can emerge out of networks of neurons, each of which is quite simple. That's our working hypothesis, anyway."

Like many others studying the patterns of these puzzling waves of neuronal excitation, Wu cannot help speculating whether they play some fundamental role in that eternal mystery: how billions of interconnected neurons, each hardly more complex than a biological light switch, can all add up to something that reasons, feels and thinks. Even if, as in the case of rats, all it thinks about is how best to gnaw its way into your food cupboard.

~

Shall we return to the three different forms of waves?

The second ones are longitudinal waves. These are any whose oscillations move forward and backward, in line with the wave's *Longitudinal, or earthworm waves* travel, rather than from side to side. So, if transverse waves are snake waves, then longitudinal waves are earthworm waves. For these little plowmen, so essential to horticulture, move through the soil by making regions of muscular contraction and expansion travel down the length of their bodies. Where a

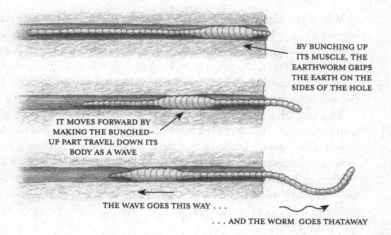

BY BUNCHING UP
ITS MUSCLE, THE
EARTHWORM GRIPS
THE EARTH ON THE
SIDES OF THE HOLE

IT MOVES FORWARD BY
MAKING THE BUNCHED-
UP PART TRAVEL DOWN ITS
BODY AS A WAVE

THE WAVE GOES THIS WAY . . .

. . . AND THE WORM GOES THATAWAY

Where would gardeners be without these little longitudinal waves?

worm tenses its muscle, the body bunches up and thickens, gripping the earth around it with the help of microscopic spiny outgrowths called "chaetae." It pushes itself forward as the region that it bunches advances in a wave down its length. The movements of its segments, as it burrows like this, don't go from side to side like the snake's serpentine locomotion, but forward and backward, along the direction it is traveling.

The earthworm's longitudinal muscular waves are therefore very different from the sideways undulations of serpentine motion. But some snakes do use longitudinal waves to get around, too. This is either because they want to creep along in stealth mode or because they are just too bulky to be able to slither from side to side over land. Confusingly, this is therefore an example of a snake using an earthworm wave.

The enormous 20ft-long African rock python is one of the fatties. It edges forward by means of subtle longitudinal ripples that travel down the length of its body. Boa constrictors, which are also of a more "generous" build, do the same thing. Known as "rectilinear locomotion," this version of earthworm movement involves the large snakes inching forward in a straight line by contracting and expanding their muscles in a sort of snake belly dance.

Where the muscles contract and bunch up, the scales along the snake's belly splay out slightly from its body. This allows it to dig them into the ground like hundreds of little fingernails, much as an earthworm uses its chaetae. By making the contractions and expansions ripple down the length of its belly, the snake edges forward as the regions gripping the ground move back.

The lack of side-to-side movement makes this a good way for some nimbler snakes to perform a "don't-mind-me-I'm-just-a-boring-old-branch" tactic in the final moments of hunting prey. Whether fatties or thinnies, snakes can perform this style of locomotion only if they have the requisite combination of powerful, well-defined muscles and loose skin. To us humans, this sounds a rather unlikely combination, almost akin to having rippling biceps and flabby arms.

The belly control required to use longitudinal waves like this makes it sound like an incredible effort, especially if you are already a snake of size. But rectilinear locomotion is in fact extremely energy efficient. The overall muscular movements are actually very subtle. A giant African python need use up only twenty calories a day getting around like this, which, in terms of calorific intake, is the equivalent of eating one raw quail's egg.* This seems a shame: it sounds as if it could do with a bit more exercise.

⌐

You may be interested to learn that your cortex is structurally rather similar to that of a rat. So if little spiral waves dance across rodent brains as they're drifting off to sleep, perhaps similar tiny electrical hurricanes form in *your* cortex as you lie there at night. The next time you're unable to sleep because you have some annoying song stuck in your head—"You're Beautiful" by James Blunt, perhaps—you just need to will the tiny depolarization waves into action. If you can stir them up a bit, encourage them to swirl around the undulating folds of your gray matter, they might just help detach

---

* Apparently, after consuming something as big as a gazelle, crocodile, or teenage child, a rock python can survive for as long as a year without another meal.

your neocortex from the stimulating control of your thalamus, thereby releasing your conscious mind from that cheesy pop song.

Exerting this degree of self-control may sound impossible, but, using a technique known as "neurofeedback," people can and now do observe the electrical activity in their brains, and even learn to control its behavior. Believe it or not, they do this by playing a computer game using nothing but their thoughts. *Look, no hands!* Imagine controlling a console with no joysticks, no buttons, no gizmos—just a pair of tiny gold sensors pressed against your scalp, which pick up the electrical signals within your brain and make the characters move on the screen. With a neurofeedback machine, you control the action by learning to change the rhythms at which your neurons fire.

But don't expect to find one under the tree this Christmas. The computer games themselves are generally quite lame, as they're designed not for pleasure but, rather, to reveal, or "feed back," the rhythmic pulses of electricity normally hidden within your head. And once you are able to see them, you can learn to influence them.

Why would you want to do that? Well, if you are unlucky enough to suffer from epilepsy or attention-deficit disorder, for starters, or perhaps if you are preparing for a particularly challenging musical performance. Oh, and should you be planning to take a penalty kick in a World Cup soccer match.

In 1924, Hans Berger, a German scientist, discovered that our brains fire in regular rhythmic pulses when he made the first human "electroencephalograph," or EEG recording. He placed silver-foil electrodes against the scalp of his fifteen-year-old son, Klaus, and measured the electrical signals from the neurons in his brain.

As one neuron triggers another, a tiny electric current crosses the gap, or synapse, between its branch and the other's cell body. Although metal-disc electrodes held against the scalp are far too crude to pick up a single neuron firing, early neuro- *"I promise this won't hurt, Klaus"* scientists like Berger found that they do pick up shifting electrical signals of a few thousandths of a volt, caused by the collective activity of thousands of neurons. These are the brain cells just below the position of the electrodes, in the outer layer of the brain, the cerebral cortex.

When he observed Klaus's brain rhythms, Berger found that, rather than showing the sort of random static you might expect from thousands of neurons whirring away, the signals had a distinct pulse to them. In fact, when Klaus was sitting in a calm but alert state, the voltages varied but the pulses themselves remained remarkably regular: always around ten "cycles," from negative to positive voltage, per second.[3]

There can't have been a dull moment in the Berger household. The scientist also stuck the EEG electrodes on the scalp of his fourteen-year-old daughter, Ilse, and told her to divide 196 by 7.

"Ilse, can I help with your math homework?" While she did the mental arithmetic, the pulsing signal sped up. I don't know whether his teenagers got fed up and told him where to stick his brain-wave machine, but he soon tried measuring the signals from young babies and toddlers. Unable to find a pulse in the youngest ones, he concluded that babies' developing brains lacked any discernible rhythm until they reached the age of least two months. He was clearly unable to restrain himself from hooking up anyone he came across. Berger even measured the signal from a dying dog and found that, as the old mutt's life ebbed away, the EEG trace flat-lined.

Berger's ten-cycles-per-second rhythm turned out to be just one of a range of EEG frequencies produced by the human brain. The dominant frequency of these "brain waves" will depend on where

A REFERENCE SIGNAL, JUST TO SHOW A REGULAR 10 PULSES PER SECOND

THE SIGNAL FROM ELECTRODES ON THE SCALP OF BERGER'S FIFTEEN-YEAR-OLD SON, KLAUS

↑ NO DOUBT, WHAT KLAUS ↑ WAS THINKING ABOUT ↑ AT THE TIME ↑

girls   *acne*   *girls*   *my dad is such a weirdo*

In 1924, when Hans Berger recorded his teenage son's brainwaves by attaching electrodes to his scalp, he demonstrated that the boy's neurons fired with a regular pulse.

the electrodes are placed and on the subject's general state of arousal: whether they're awake or asleep, eyes open or closed, concentrating on something mentally taxing or watching *American Idol*. Scientists have divided the frequencies into four different bands.

~C~

Known as delta waves, the lowest frequency brain waves are 4 cycles or fewer per second. Most delta-wave activity happens during deep sleep, unless you are a baby, in which case it is the dominant frequency when you are awake, too. Delta waves are also sometimes observed in coma patients. In the 4–7 cycles-per-second range are theta waves, which are frequencies most common when you are drifting off to sleep. Theta are the most embarrassing brain waves, since this is the frequency range most commonly associ- *A brain* ated with an unsightly dribble of saliva dangling from your *wave for every* bottom lip as your head droops on the morning commute *occasion* to work. Alpha waves, ranging from 8–12 cycles per second, dominate when you are calm and relaxed. Anywhere above 12 cycles per second, and you are in beta-wave territory. Fifteen to eighteen will be the dominant frequency when you're concentrating on something complicated, like this sentence.*

In the 1970s, Dr. Barry Sterman of the UCLA School of Medicine showed that epileptics who learned to change the rhythm of activity in a particular region of their brains, across the top of their heads, could significantly reduce the number of seizures they experienced.[4,5,6] During an epileptic fit, a patient's brain waves behave in an abnormal manner. Although there are many types of seizure, they are often characterized by powerful EEG voltages sweeping across the whole brain as pulses synchronized across all areas. This couldn't be more at odds with the usual brain activity, when the various regions work at different frequencies, as they perform their separate processing tasks. An epileptic seizure is therefore much like

---

* Some neuroscientists describe high beta frequencies, of 40 cycles per second or more, as gamma waves. Bursts of these are sometimes recorded during REM (rapid eye movement) sleep and meditative states.

a tidal wave of electrical activity sweeping across the brain. As the synchronized pulsing tends, in adults, to be in the theta range of 4–7 cycles per second, Sterman used neurofeedback to train patients to try to stop these synchronized theta waves from occurring.

He placed his EEG electrodes over an area of the brain known as the "sensorimotor strip," which lies beneath the top of the patient's head, and is associated with muscle control. Most people, when they are actively relaxing their muscles, produce bursts of brain-wave activity in this part of the brain at a rate of 12–15 cycles per second. This particular frequency range, low beta, is so characteristic of this region during relaxation that it has become known as the "sensorimotor rhythm," or SMR. Sterman's logic went like this: if controlled muscles are associated with 12–15 cycles per second in this region and epileptic seizures are with 4–7 cycles everywhere, patients might be trained to produce more of one frequency and less of the other. Sterman taught epileptics to control their brain waves with a device that flashed a green light when the frequencies in their sensorimotor strips were in the SMR range, and a red light when they dropped to the theta range.

With this training, the patients learned to strengthen the brain waves associated with muscle control, although they found it hard to explain exactly how they learned to change the brain rhythms. To make the green light come on, Sterman explains, requires a sort of active relaxation, a concentration on calming the body. "It's when we will ourselves to be still. It's a standby state for the motor system. You might think of it as a pause button."[7] By learning, session after session, to make the green light come on and the red one stay off, the epileptic patients did indeed learn to increase these SMR frequencies and to inhibit their theta frequencies. And this led to a dramatic improvement in their condition.

The effectiveness of neurofeedback training on epileptics has been demonstrated time and again.[8,9,10] In 2000, Sterman conducted a review of all the research worldwide into neurofeedback on epileptics and found that the studies had reported "overwhelmingly positive results." Eight out of ten patients who were not taking drugs for their epilepsy had shown at least a 50 percent reduction in the frequency of their seizures following neurofeedback training. Five percent had remained completely seizure-free for up to a year *after* the treatment finished.[11] Neurofeedback has now been established as a well-founded and viable alternative to medication.[12]

The technique is also used to treat neurological conditions such as attention deficit and hyperactive disorders (ADHD) in children.[13] But keeping children with attention problems focused is easier said than done, so new neurofeedback programs were designed in which the children's brain waves are fed back to them not with green and red lights, but with computer games. When they increase the desired frequency of brain waves, they progress to the game's next level; when they revert to the problematic frequency, they're held back.

"A majority of children with ADHD have too much slow, theta frequency appearing at the front of their brain, compared with the faster, beta frequencies," Melissa Foks, a London-based neurofeedback practitioner, explained to me. But aren't the theta waves the ones you produce as you drift off? Surely they're the one type of brain wave that hyperactive children could do with more of?

"Imagine you're driving on the motorway," Foks explained. "It's late at night and you're desperately trying to stay awake. You might wind down the window, turn up the radio really loud, sing at the top of your voice. You'll do anything you can to try and keep yourself awake." In the same way, a hyperactive child *Attack of the theta waves* is battling against the theta-frequency brain waves that are associated with doziness. This is why patients suffering from ADHD are often prescribed *stimulant* medication, such as Ritalin, which has the paradoxical effect of calming them down. "They are going crazy trying to keep themselves from drifting," added Foks, "and it's pretty unhelpful behavior in a classroom situation."

While the benefit of neurofeedback training for ADHD and epilepsy has now been demonstrated in rigorous clinical studies, there are many more conditions that it is used to treat—autistic spectrum disorder, head injuries, drug addiction, and depression among them. The evidence for its effectiveness as a treatment for these problems tends to be based more on individual case studies, and so is less robust scientifically.

But neurofeedback is not just used to treat brain disorders.

Several of the 2006 World Cup-winning Italian football team underwent this training to help them keep their nerve in penalty shoot-outs. Of course, without anything to compare with, no one can say for sure if the neurofeedback had any beneficial effects.

Not so in the case of students at the Royal College of Music in London, who had neurofeedback training to reduce nerves before a musical performance.[14] They were trained to reduce the amount of faster, alpha frequencies and increase the amount of slower, theta ones. The logic was that increasing theta, the dozing-off-on-the-train ones, would help the students stay relaxed in their performances.

They were each filmed performing the same pieces before and after the ten sessions of treatment. These were then judged by a board of external examiners. The videos were jumbled up so that the examiners didn't know which was before and which after, and they *Musical* were also mixed in with the performance films of students *guinea pigs* who had undergone various alternative treatments to help relaxation. These other students had spent the same time doing physical exercise, mental skills training, the Alexander Technique (a posture training used to reduce tension), or neurofeedback that emphasized different frequencies of brain wave (to account for any placebo effects of such an exotic-seeming treatment).

When the examiners marked the performances, the results were astonishing. Without having any idea which students had undergone which training, nor which performance was "before" and which "after," they judged the students who'd received neurofeedback that increased their "relaxed" theta waves to have improved their musical performances by the equivalent, on average, of two years' experience. None of the other students were considered to have improved at all. They must have felt a bit short-changed—except the physical training ones, who would at least have got fit.

~

If all these brain waves feel overly cerebral, then how about returning to something more tangible? Like the last of our three mechanical waves: torsional ones.

Where transverse waves consist of side-to-side movements and longitudinal waves of forward-and-back, torsional waves travel by means of a twisting motion. Such movements can be rather subtle, so it is not a type of wave you are likely to have noticed. Torsional waves are able to travel along anything that resists twisting by springing back to where it started. Say you cemented one end of a long, metal rod into a wall so that it stuck out at a right angle and then you welded a steering wheel onto the other end and used this to give it a good twist before letting go. Torsional waves would travel up and down the length of the rod between the fixed end and the steering wheel, twisting from side to side. What do you mean, you can't imagine ever doing that?

Workers in the drilling industry are about the only people who spend any time thinking about these waves. The shifting stresses caused by boring into rock send torsional waves racing up and down the rods and strings of the drilling equipment. So definitely bear that in mind when you are next designing an oil rig. The rest of the time, you needn't overconcern yourself with torsional waves, for they are far less common than their fellows.

Which presents a bit of a problem.

If I am to complete the trilogy of wave-motion demonstrations using animal movements, I'll need to find an example of some little critter that uses torsional waves as a means of locomotion. The trouble is that I can't, for the life of me, find a single one that does.

The only creatures that come close aren't even animals, in the strict sense of the word. Instead, they are microorganisms, among which one finds certain types of bacteria propelling themselves along using tail-like flagella, resembling those behind human spermatozoa. The *E. coli* and *Salmonella* bacteria are examples. Some strains not only propel themselves along very efficiently with their wiggling tails but, when eaten in contaminated food, also propel you to the bathroom rather quickly, and often on to the hospital. But these microbes don't wiggle their flagella from side to side like the sperm do, thereby producing straightforward transverse waves. Instead, their tails spin around on tiny nanoscale motors, somewhat like pieces of rope attached to boat propellers. As they twist, the flagella flail in all directions, propelling the cells forward.

These bacteria seemed perfect. The only hitch was that there is no evidence that torsional waves really do travel down the length of their flagella. The steady rotation is just a good way to make the tail flap up and down and side to side at the same time. These are just 3D transverse waves masquerading as torsional ones.

Damn. Since I can't find any examples of creatures using torsional waves to move around, can I get away with an instance of a torsional wave *stopping* one in its tracks?

I should warn you that this is a sad story. It doesn't end well for the animal in question: a three-legged cocker spaniel called Tubby.

꿎

Tubby's encounter with torsional waves took place in 1940, while he was traveling in the back of a car being driven over a bridge near the town of Tacoma on Puget Sound, some 40 miles south of Seattle, Washington. Leonard Coatsworth, a local journalist, was at the wheel, while Tubby—his daughter's pride and joy and a beloved family pet—was in the back.

Torsional, or Tubby waves

The Tacoma Narrows Bridge had been wobbling since it was opened just four months earlier. Indeed, it had vibrated so much during its construction that its builders had christened it "Galloping Gertie" and many workmen had taken to chewing lemons as they worked to counter the motion sickness. But, until that day, these vibrations had been no more than rolling undulations, in which the whole span of the bridge lifted gently up and down in the wind.

The Toll Bridge Authority had hired a certain Professor F. B. Farquharson, from the engineering department at Washington State University, to work out how to dampen these transverse waves. No one considered them to be dangerous, and the bridge was used as normal, in spite of the wobble.

But on November 7, a steady 42mph wind had caused the central span of the half-mile-long suspension bridge to develop an alarming twisting motion. This had become so violent by the time that Leonard Coatsworth had reached the middle that he was unable to keep control of his car and had to slam on the brakes. The concrete around the car had also begun to crack and as he

Professor Farquharson returns from trying to save Tubby on the Tacoma Narrows Bridge. Can you spot the subtle movements of the torsional waves?

jumped out he was thrown onto the tarmac. Unable to get to the door to rescue poor Tubby, Coatsworth crawled the 500 yards to the stability of the bridge tower, his hands and knees raw and bleeding from being thrown about on the twisting surface.

Clearly some new, and more violent, mode of vibration had been set up by the wind that day. As soon as he had heard news that the bridge was "galloping" more than ever, Professor Farquharson had grabbed his movie camera and driven down to take a look. When he arrived, he could see instantly that the vibrations along the span of the bridge had the twisting form of torsional waves, rather than the familiar up-and-down transverse ones. That morning, the carriageway of the central span lifted first on one side and then the other, as twisting waves traveled down its length and back. The wind was blowing at just the right speed to cancel out the structure's tendency to settle down from the twisting motion. So, as the span began to flutter slightly in the steady wind, the twisting vibrations gradually grew more and more violent.

Setting up his camera at one of the bridge's piers, where the twisting motion was minimal, Farquharson filmed the car belonging to Leonard Coatsworth swaying and lurching in the central span.[15] Coatsworth had told him of the poor dog cowering in the back. The professor must have been a dog person rather than a cat person, for he decided to do the right thing and headed out to save it.

Pipe in hand, Farquharson walked gingerly out onto the central span from his camera at the tower. He followed the road marking in the middle of the roadway, as this was the axis along which the bridge span was twisting, and so was relatively steady. But the edges were now rising and sinking by about 10ft every couple of seconds.

Leaving the security of the central road line to reach the car on the left-hand side of the bridge, Farquharson staggered across like some drunken stunt man and opened the back door to try and coax Tubby out. But the lurching, swaying motion of the torsional waves had so terrified the poor dog that he instinctively bit the professor's hand—who, himself struggling with the bridge's violent movements, decided to leave Tubby where he was. That bite was to cost the spaniel his life.

Not long after the professor had staggered back to the safety of solid ground, the bridge's central span collapsed. The torsional waves had finally weakened the girders so much that they went crashing into the waters below, taking the car, and poor Tubby, with them.

In newspaper reports the next day, Leonard Coatsworth recalled the moment that he watched the bridge fall: "With real tragedy, disaster and blasted dreams all around me, I believe that right at this minute what appalls me most is that within a few hours I must tell my daughter that her dog is dead, when I might have saved him."[16]

~

While we can divide physical waves into transverse, longitudinal, and torsional waves—into snake, earthworm, and Tubby waves—the truth is that many real waves combine aspects of more than one type. Take ocean waves out at sea, for instance: those circular paths, along which the water moves as the wave passes through, are a combination of up-and-down and forward-and-back motion. The water might look like it just rises and falls with the passing wave, but it actually combines transverse (up and down) with longitudinal (forward and back) movement, resulting in orbital motion. You can feel this when swimming in deep water. Not only do you have the sensation of rising and falling with the passage of the wave, you also notice that you are sucked toward the crest as it approaches you, and dragged along behind it momentarily as it rolls on by. In the approach to shore, the orbits of the water below the surface become increasingly flattened, as the transverse movement is constricted.

After that awkward business with the torsional waves, you'd think the chances of finding an animal using a combination of transverse and longitudinal to propel itself forward must be zero. But the perfect animals do exist: gastropods or, as you and I know them, slugs and snails.

Now we're back on track, even if it is a slimy one.

These villains of the vegetable plot cunningly combine transverse and longitudinal movements as they slither through the dead of night toward the stalk of your prize broccoli. You have to watch the soles of their glistening feet with care to see what is going on. But their undulations are so subtle that you'll stand no chance unless you observe their gooey waves when they are climbing up the pane of a window. Watching from the other side, you *Multitasking* will see tiny lighter and darker ripples of muscle traveling *gastropods* along its length. The darker bands show where the sole is lifted slightly from the slime trail, and the lighter where it is pressing against it. What is surprising is that, in the case of the common or garden slug or snail, the muscular waves travel from the tail back toward the head. The gastropod lifts its tail slightly, bunches it up and then places the tip down on the glass slightly farther forward than before. The subtle kink in the sole, so formed, then progresses forward until, at the other end, the tip of the head is lifted and placed down a little way forward.

In fact, several such waves travel forward at a time, the tail lifting and pressing down farther forward to start another wave just as soon as the previous one has started up the sole. Some gastropods

THE FOOT
THE SLIME
THE GLASS

THE RIPPLES ON THE SNAIL'S
FOOT TRAVEL FROM ITS TAIL
TO ITS NECK 20 TIMES AS
FAST AS THE SNAIL INCHES
FORWARD OVER THE GLASS

EACH PART OF THE SNAIL'S
SOLE MOVES IN TINY,
ELONGATED ORBITS AS IT
LIFTS FROM AND PRESSES
ONTO THE LAYER OF SLIME

Transverse and longitudinal waves are subtly combined by this slimy visitor,
and are visible as it slithers up your window pane.

move using waves that travel from head to tail and some even produce separate waves down either side of their soles that are out of sync with each other, lifting on one side while pressing down on the other. Regardless of which direction the waves travel along the foot, each part of the muscle from side-on moves in roughly oval orbits. The waves pass along its length by every point on its surface, moving forward and back as well as up and down. I think we can safely say that those oval orbits are the one and only similarity between slugs and ocean waves.

～

"Nature only knows one thing," Saul Bellow wrote in *Seize the Day*, "and that's the present, present, present, like a big, huge, giant wave—colossal, bright and beautiful, full of life and death, climbing into the sky, standing in the seas."

The waves within us seem such a fundamental internal transport system for our bodies that I can't help wondering what happens to them when we die. Do they "break," like ocean waves at the shore?

The involuntary muscular waves that transport food, blood, and more through our systems, and the electrochemical waves that move information through our nerves and brains, differ from ocean waves in one essential regard: none of them are self-sustaining. Waves on the surface of the water, once they have been set in motion by the wind, can travel for some distance over the surface due to the force of gravity and the water's surface tension, without the need to be pushed along by the wind. The waves within our bodies move because energy is being constantly fed in. Every beat of our heart uses up energy. Every firing neuron burns up calories. The waves traveling through our muscle and nerve tissues will not roll on once the breath of life is exhausted. When we die, they merely stop in their tracks. The reactions that powered their progress have been shut down.

Nevertheless, it is hard to relinquish the feeling that our bodies are energized in a way that echoes the animation of the water by a wave's energy. When it breaks upon the shore, the wave's energy dissipates into the surroundings, for energy never disappears; it only changes. And so the energy that sustains us ebbs away into our surroundings when the chemical engines of life grind to a halt.

Who knows where the waves that course through us day and night eventually break? No one can chart the foreign shores upon which they finally tumble after death. As Thomas Hood, the now unfashionable nineteenth-century English poet, wrote:

> We watch'd her breathing thro' the night,
>   Her breathing soft and low,
> As in her breast the wave of life
>   Kept heaving to and fro. . . .
>
> But when the morn came dim and sad
>   And chill with early showers,
> Her quiet eyelids closed—she had
>   Another morn than ours.[17]

The Second Wave

WHICH FILLS OUR WORLD WITH MUSIC

As wrote the poet Oliver Wendell Holmes, "A word, whatever tone it wear, is but a trembling wave of air."[1]

A spoken word, or any other sound, is an audible "acoustic wave." I say audible because, ridiculous as it sounds, most acoustic waves can't be heard.

The term acoustic actually refers to *any* sort of wave that travels through "stuff"—be it solid, liquid, or gas—by means of the stuff being compressed and expanded. They are quite different from waves on the surface of the water. When the "crest" of an acoustic wave arrives, the material it is traveling through bunches up into a passing region of compression, or increased density. In other words, the medium doesn't lift and sink, as with sea waves, it only moves backward and forward. For this reason, acoustic waves are longitudinal ones. The physical movement of the medium is like the muscles of an earthworm, expanding and contracting as it inches through the soil. When the "trough" of an acoustic wave arrives, the material it is traveling through expands into a passing region of "rarefaction," where the material becomes slightly less dense than normal.

So what is the difference between an acoustic wave that you can hear and one that you can't? The answer is that we hear them only when the pattern that reaches the air in our ears takes the form of a *succession* of compressions and rarefactions, when it is a "periodic" wave. A single rise and fall in the air pressure—a solitary crest of a longitudinal wave—doesn't generally register as a sound.* But there is another essential requirement for us to be able to hear an acoustic wave: it needs to be repeating at the right rate. Pressure waves passing through the air in our ears have to make our eardrums vibrate at between 20 and 20,000 times a second for us to hear them. The slower vibrations, or low frequencies, we hear as low sounds; the faster vibrations, or high frequencies, we hear as high sounds. Outside of this range, talk to the hand, 'cause the ear ain't listening.

Only a small fraction of all the acoustic waves flowing through our environments are actually apparent to us. Most traveling patterns of compression and rarefaction don't register. Just *Inaudible* beyond the range of our hearing are the low "infrasound" *acoustic waves* frequencies, such as those produced by elephants making long-distance calls, and the high "ultrasound" frequencies, such as those used by bats and porpoises to echolocate. These acoustic waves still vibrate our eardrums but don't sound like anything to us. There is nothing inherently audible about acoustic waves.

You may make no sound as you wave your hand from side to side in saying goodbye to someone, but you are, nevertheless, producing acoustic waves. By sweeping this way and that, your hand compresses and expands the air on either side. These local-ized differences in the air pressure spread outward as acoustic waves. We don't call them sound waves because we can't hear someone waving. This fact proved rather unhelpful for the poor

---

* If it is a particularly sudden change in air pressure, we might hear it as a click, like the sound of an electric spark. If it is a particularly large change in pressure, we are more likely to feel it as a popping of our eardrums.

soul in Stevie Smith's poem about a man drowning out in the surf, unable to catch the attention of those on the beach:

Nobody heard him, the dead man,
But still he lay moaning:
I was much further out than you thought
And not waving but drowning.[2]

You might think that no one hears you wave because the sweep of your hand doesn't change the air pressure enough to vibrate anyone's eardrums. But we actually have very sensitive little drums in our ears. If the pressure changes flowing into our ear canals fluctuate at the right rate, we can hear them as sound even *Waves made* when the air pressure shifts up and down by no more than *by waves* 0.01 percent. In fact, the silence of your hand waving is not due to the amount of compression and rarefaction of the air, but to the speed of the pressure shifting. The reason no one hears it is simply because you aren't flapping fast enough.

Compare the huge movements of a hand waving back and forth, for instance, with the tiny ones of a bee's wings. You can easily hear the buzz of a bee flying in to join your picnic because it flaps its wings at a rate of about 180 times a second. This is described as a frequency of 180 "hertz," after the nineteenth-century German physicist Heinrich Hertz, who first demonstrated the existence of radio waves.

Once the bee returns to the hive to tell his friends about the exact location of my custard tart, it does so by wiggling its abdomen as it crawls around among its bee friends. The movements are even tinier than their wing flaps, and also tend to be faster—more like 500 hertz.[3] But the acoustic waves produced by these minuscule pressure fluctuations are completely audible, as they fall well within the 20–20,000-hertz range of our hearing. The wing-flapping and body-wiggling produce air-pressure beats with quite steady rates, so we even hear them as notes. For the more regular the succession of pressure crests reaching our ear, the more they sound to us like a pure note. A beekeeper with perfect pitch might know that the higher buzzing sound, when a bee wiggles its abdomen to tell the

F-sharp below middle C.

others the direction of the beeline to your picnic, is equivalent to the note B, which seems rather appropriate. It is the B above middle C on a piano. The deeper buzz, as the blighters flap their wings while flying to join the feast, is more like an F-sharp below middle C. It is just as well that our eardrums are so sensitive. The total power transmitted in the form of sound waves by a large orchestra playing at full-tilt amounts to no more than the power consumed by a single 100W lightbulb.[4] It is important to emphasize that it isn't the air itself that travels from the orchestra to our ears. This remains largely in place, while the energy reaches us as localized vibrations of the air. The music is not a Mozart wind.

Our hearing systems are also incredibly sophisticated, for the sound waves emanating from each and every instrument merge and coalesce as they cross the concert hall to reach us. They have to, *A cough from* since they all animate the same air—air that can, of course, *the string section* only be compressed or expanded by a single amount at any place and at any time. So the waves combine into a single pattern of vibrations, a single, complex succession of compressions and rarefactions that coax our eardrums to move in sympathy with them. Is it not truly miraculous that our brains can untangle such a chaotic succession of tremors—that we can decode the microscopic movements of this stretch of skin, just ¼ in across and 2 thousandths of an inch thick, with such exquisite acuity that we can notice the second violinist cough halfway through the second movement?

~e~

That we may not be able to see that sound is a wave doesn't make it any less wavy. In fact, sound waves exhibit classic wave behavior, so they elegantly demonstrate the three means by which waves change direction. These are known as "reflection," "refraction," and "diffraction."

They might sound like terms thought up by stuffy old physics teachers—in fact, they probably *are* terms thought up by stuffy old physics teachers—but don't let that turn you off. These ubiquitous wave characteristics are at the heart of how we perceive the world. Once you understand them, you are well on your way to wave-watching enlightenment.

In my own family, reflection, refraction, and diffraction have come to be treated with reverence, and are now known simply as "The Ways of the Wave."

Let's start with reflection. The First Way of the Wave is simply:

𝔚𝔞𝔳𝔢𝔰 bounce off stuff.

Yes, I know. Hardly the revelation of the century. But it turns out that waves don't just bounce like balls do. They rebound with far greater sophistication than anything you could manage to make a ball do on a squash court.

Incidentally, it is worth noticing that the way sound bounces off walls to produce echoes was what first drew thinkers to consider that sound might share similarities with the waves you see on the water. For instance, when writing toward the end of the first century BC about the need to take into account sound reflections in the design of theaters, the Roman architect Marcus Vitruvius Pollio (often known as plain Vitruvius) proposed that "voice is a flowing breath of air," which is

> *perceptible to the hearing by contact. It moves in an endless number of circular rounds, like the innumerably increasing circular waves which appear when a stone is thrown into smooth water.*[5]

Since ripples are seen to reflect off the walls of baths and head back the way they came, it seems quite sensible to wonder whether sound bounces off the walls of theaters because it is also a wave. Of course, Vitruvius's description of sound as a "breath" of moving air was wrong in the same way as the Mozart wind was. When someone shouts at you angrily from across the room, you may feel a blast of fury but you do not experience a gust of wind. Sound waves travel *through* the air with no overall displacement of it. Nevertheless, Vitruvius's analogy between sound and water waves was inspired. Since it is generally invisible to us, the waviness of sound must be gleaned from its actions, rather than its appearance.

∿

The seventeenth-century German Jesuit and polymath Athanasius Kircher was interested in echoes. Capable of speaking dozens of languages, including Chinese and Coptic, Kircher's enormous body of work included books on geology, optics, astronomy, and

Athanasius Kircher's drawing, published in 1673, to show how the delay of echoes depends on the distance of a reflecting wall.

acoustics. By rights he should be the patron saint of union leaders since he also invented the megaphone.

In *Phonurgia Nova*, a book he published in 1673, Kircher described an experiment in which someone stands in front of a series of partitions sticking out from a wall and calls out a word. He included a diagram, showing these partitions protruding at right angles from the wall, and at different distances from the caller. As the sound reflected off each successive partition, the echo returned with a greater and greater delay. By yelling the Italian word *clamore*, which means "shout," Kircher realized you would hear a succession of shortening fragmentary echoes: *–lamore, –amore, –ore, –re*. This is because each successive echo, being a little more delayed than the last, returns to you with less of it overlapping you as you complete the word. And each fragment happens to have a meaning of its own in Italian: "costumes," "love," "hours," and "king."

Quite what significance can be drawn from the wordplay is not clear, but I'm sure that the challenge of coming up with words whose partial echoes have their own meanings could take off as a parlor game—that's assuming your parlor is an enormous cave.

For within such a cave, whatever phrase you utter will be bounced back at you, but it is only when the walls are far enough away that you notice it returning after you've finished speaking. When the dimensions are smaller, the sound overlaps your utterance so fully that you don't hear it as an echo. The near-simultaneous reflection of sound in this situation is known as "reverberation," and is part of what we describe as the acoustics of a space. We derive all sorts of cues about our surroundings from the quality that the reflections give to the sound.

∿

It seemed as if our daughter learned to speak through echoes. At one and a half she tended to repeat whatever we said. If I greeted her with exaggerated formality at the entrance to her Wendy house, saying, "You must be Flora. I've heard a lot about you. It is very nice to meet you." She would gleefully shake my hand and then reply, ". . . Meet you."

For a period she seemed the incarnation of Echo, the nymph from classical Greek mythology who, in the words of Ovid, "cannot be silent when others have spoken, nor learn how to speak first herself."[6] It was Hera, the ever-frustrated wife of Zeus, who bestowed upon Echo this rather unfortunate speech impediment.

Hera's revenge Understandably, Hera had become pissed off by Echo's habit of covering up for all the other nymphs who were having affairs with Zeus behind her back. Given Echo's technique—which was to engage the goddess in lengthy conversation as her fellow nymphs made their escape—Hera's punishment was to rob Echo of the power of original speech so that she could only ever repeat the last of whatever words had just been spoken to her. Overnight, her nymph friends must have found that she'd turned into one of those annoying people who try . . . to finish . . . your . . . sentences for you.

Unable to strike up a conversation with Narcissus, the unfeasibly beautiful youth for whom she had the hots, Echo could only slink around in the shadows after him. She became more and more flushed with desire, desperately hoping that one day when they were alone he would finally utter something that she could repeat.

Her opportunity eventually arrived when Narcissus, having become separated from his friends, called out, "Is anyone there?"

Echo seized her chance: " . . . anyone there?" she teasingly repeated. And it turned out that repeating his words worked a treat. Narcissus was intrigued as to who owned this mysterious voice.

"Here, let us meet together," he called out.

" . . . together," she simply replied, delighted with how things were shaping up.

But then she made a classic courtship error: she lunged too soon.

Running out of her hiding place in the woods, Echo threw herself upon Narcissus, draping her arms around his neck. As Ovid describes it, this didn't go down well:

> He runs from her, and running cries "Away with these encircling hands! May I die before what's mine is yours." She answers only, "What's mine is yours!"[7]

Poor Echo. What a basic error: one that is normally the preserve of spotty male teenagers at their first parties. She took it all so badly that she withered away, leaving behind nothing but her voice. Frankly, I'm not surprised. It makes me wince even to write it down.

❧

Did I promise to explain why the reflection of waves is a far more sophisticated process than the bouncing of a ball? Well, the answer is that, as they reflect, waves divide. Since they consist of energy moving from one place to another, it is not unusual for waves to be split so that some of the energy heads this way and some that. Dividing into smaller waves is something that comes easily to them, and it is something they do whenever they bounce—sorry, *reflect*—off something.

Actually, they don't just bounce

When a wave traveling through one medium hits a boundary with a differing medium, one that is made of significantly different stuff, some but not all of its energy is reflected. A portion of it keeps on going beyond the boundary, is "transmitted" through it, to carry on through the different medium. This is not something that happens when a football bounces off a goal post. Sure, the ball dissipates some of its energy into the post as it ricochets off, but there is no sense in which part of the ball bounces back to the field and part of it goes on through the goal post. I wouldn't want to be the referee if there were. You can see this part reflection/part absorption thing happening all the time with sound waves—or, rather, you can hear it.

Say that your other half is boring you as you sit in the bath by recounting every tedious detail of his or her day at the office. You might feel an urge to sink your head below the surface of the water for a few moments. If you do so, you will notice that, while much of the tedious talk is reflected away, not all of it is. A sort of low-frequency drone of speech, which lacks the hiss of the sibilant sounds, passes down through the water and still vibrates your eardrums. So in spite of the protective bathwater, you will still hear muffled murmurings about how the head of marketing is getting on everyone's nerves.

The partial reflection, such a common feature of sound waves, has serious military implications. The active SONAR system used by one submarine to find the direction and range of another sends out a "ping" of sound waves and listens for the echoes bouncing back. The direction the echo comes from, and the length of the delay as the waves travel there and back, can indicate where the enemy sub is lurking. The only problem is that not all of the sound energy bounces off the enemy's hull. Some of it is not reflected at the boundary between water and steel and passes into the metal structure. With the right microphones, the enemy can always hear the ping and know that someone is in the vicinity. This is why military submarines tend to have their SONAR systems turned off half the time.

The process has other, more peaceful, applications, too. For instance, this characteristic of partial reflection and transmission is

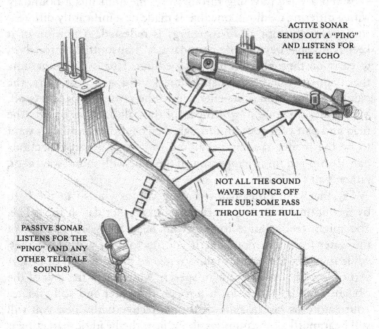

ACTIVE SONAR
SENDS OUT A "PING"
AND LISTENS FOR
THE ECHO

NOT ALL THE SOUND
WAVES BOUNCE OFF
THE SUB; SOME PASS
THROUGH THE HULL

PASSIVE SONAR
LISTENS FOR THE
"PING" (AND ANY
OTHER TELLTALE
SOUNDS)

They don't really use those sorts of microphones.

why waves are a great way to look inside your body when you go for an ultrasound scan.

The boundaries between different types of soft body tissue are nowhere near as pronounced as that between bone and flesh, but they are nevertheless enough to reflect back a portion of any ultrasonic waves passing through, and this is what produces *Reflections within the body* the ultrasound picture. Like a high-pitched, mini-SONAR, the ultrasound scanner produces pulses of sound hundreds of times higher in frequency than those we can hear, while listening for the echoes. Each boundary at which the body tissue changes in density reflects some of the wave back as an echo.

The length of time between the two indicates how far the boundary is, while the intensity of the sound wave indicates how pronounced it is, which is to say whether between muscle and bone or one soft organ and another. That part of the wave also keeps on going through any boundary allows the sonographer to "see" at different depths into the body. The reflection at one boundary is used to form an image at one depth, while the waves that pass through can form an image by reflecting off boundaries below.

The probe that produces the sound and listens for the echo has a rubbery front that is similar in density to the fatty layer covering your body. This means that, when it is held against the skin, and a liquid gel is squidged between the two, as little as possible of the sound-wave energy is reflected back before it enters the body. The change in the density is minimal as the waves pass from probe, to gel, to fat.

By stating that "waves bounce off stuff," you might claim that The First Way of the Wave doesn't really articulate the subtle ways that reflecting waves differ from bouncing balls. And you would be right. Wave watchers are thinkers, and contemplating how waves "bounce" will help them realize a fundamental truth: waves are energy passing through things, and not things in and of themselves.

Refraction, the Second Way of the Wave, is this:

Waves tend to change direction as they pass from one material to another.

Yes, I know—it sounds almost more obvious than the fact that they bounce off things.

But one fundamental aspect of waves is that they can change direction like this. In fact, refracting is a trick that every respectable wave has up its billowing sleeve.

For a wave to change direction in this way, two conditions must be met: first, it has to approach the boundary at an angle, not head-on, and second, it needs to have a different natural speed of progress through one material compared with the other. If it passes at an angle into a material through which it travels more slowly, a wave will change its bearing slightly. And if it passes into one through which it travels faster, it will shift direction the other way.

Sound performs this direction-changing routine all the time, albeit largely unnoticed. The speed of sound can vary greatly depending on what material it travels through. This might seem surprising, given the way everyone talks about "the speed of sound," as if it has a fixed speed: around the 740mph that U.S. test pilot Chuck Yeager first achieved in his X-1 jet aircraft in 1947. But the speed of sound certainly isn't fixed.

Through air, for instance, it varies greatly with the temperature. It travels at 740mph when the air temperature is 32°F. But at a room temperature of 74°F, the speed of the newscaster's voice from the TV to your ear is nearer to 770mph. This is because, for any given volume of air, the speed with which pressure changes are transferred from one region to the next depends on how fast the air molecules are moving. And the warmer any gas is, the faster its molecules move.

Sound also travels faster through liquids than through gases. This seems rather counterintuitive since you might think that water would present more resistance than air. But resistance like this applies only when something separate is pushing through a material, displacing it as it goes. This is not at all the case with sound waves. When sound moves through a material, it does so by means of the material itself vibrating. The molecules over here collide with their neighbors there, which in turn collide with the molecules further on still. Since a liquid's molecules are more densely packed than a gas, the vibrations that make up the sound waves pass through it more quickly.

Through seawater at 77°F, for instance, sound waves travel at about 3,430mph, over four times as fast as through air. At a warmer temperature, they will travel faster still. This is why precise measurements of the time taken for a test sound to travel *Sound with* under water from a speaker on one side of the ocean to *its foot on the* a microphone on the other give scientists an indication *accelerator* of ocean temperature changes from one decade to the next. The speed of sound through any liquid also depends on how dense the liquid is and how resistant it is to being compressed.

Most solids have rigid bonds between their molecules, making them even less compressible than liquids. These rigid bonds transfer pressure changes from one part of the material to the next even more quickly than the ones in liquids. And the more rigid the structure, the faster the sound waves go: 7,250mph through gold and a whopping 26,843mph through diamond.

So what do all these fancy speeds have to do with the way sound waves "change direction as they pass from one material to another"? Well, they change direction because they change speed. If a wave approaches the boundary between differing materials at an angle, then the end reaching the boundary first will change speed on entering the new material before the rest of the wave starts to do so. It might speed up or it might slow down. Either way, this has the effect of pulling the wave around to point a different way.

To use a random analogy, let's say that a UFO crashes in the middle of the desert and a group of aliens stumbles out from the wreckage, heading off in search of a McDonald's. Of course, they are a bit dazed from the impact and, what's more, they can't see very well because they are used to different wavelengths of light on their home planet. So they hold hands, or suckers, or whatever else they have, and shuffle along in a line, hoping not to attract attention.

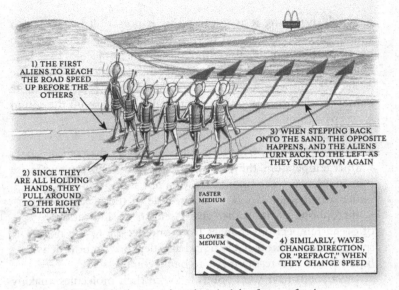

1) THE FIRST ALIENS TO REACH THE ROAD SPEED UP BEFORE THE OTHERS

2) SINCE THEY ARE ALL HOLDING HANDS, THEY PULL AROUND TO THE RIGHT SLIGHTLY

3) WHEN STEPPING BACK ONTO THE SAND, THE OPPOSITE HAPPENS, AND THE ALIENS TURN BACK TO THE LEFT AS THEY SLOW DOWN AGAIN

FASTER MEDIUM

SLOWER MEDIUM

4) SIMILARLY, WAVES CHANGE DIRECTION, OR "REFRACT," WHEN THEY CHANGE SPEED

Aliens in the desert teach us the principle of wave refraction.

It is easier for their alien feet to gain purchase on hard ground, compared with the shifting desert sands, so when they finally reach a road, the first to step onto it speeds up slightly. Since they reach the tarmac at an oblique angle, those who step onto it first start to move forward faster before the others do. Since they are all holding suckers, this has the effect of pulling the line around a little so that, once they are all on the tarmac, they are all walking along in a slightly different direction. Of course, they don't notice, as they are all too busy bickering about who hadn't been paying attention at the joystick. When they stumble back off the road again, the opposite happens: the end of the line that first steps onto the sand and slows down starts to pull the line of aliens in a different direction again. Once they are all back on the sand, they continue on their original compass bearing.

This is the principle of refraction. In the case of sound waves, the advancing region of higher pressure, or the wave front, behaves a little like the bickering aliens. When it reaches a boundary with a material through which it travels more slowly, the wave front changes direction just like the aliens. Assuming that it approaches the boundary at an angle, rather than head-on, one end of the wave slows before the rest of the wave reaches the boundary, and begins to pull the wave front around slightly to point in a different direction. So once the wave crest has passed the boundary, it is heading off on a slightly different bearing. This direction shift occurs the other way around when the sound wave enters a faster medium. Of course, sound waves are not objects moving through the air or over the ground like stumbling aliens. They are patterns of higher and lower pressure caused by oscillations of the medium. They don't have silver jumpsuits either.

Have you noticed how sounds seem to travel farther when it is foggy? It can seem that, through the fog, you can hear a laugh or the peal of church bells that are normally too far away to pick up. I've always loved this effect, as I find it adds to the sense *Church* of otherworldliness I feel when walking through fog. The *bells in fog* effect isn't, actually, due to the fog itself. The droplets of water suspended in the air are far too small to have any noticeable effect on the sound. It is the temperature of the air around ground level that causes the fog to form, and this is also what makes the peal of the church bells travel farther along the ground.

The way the air temperature changes with elevation has the effect of refracting sound waves, since they travel faster through warmer air than through colder. Normally, temperatures tend to decrease as you rise. This causes sound waves to refract upward, bending away from the ground. Since the sound from church bells tends to bend upward, they eventually lift away from the ground so much that they can no longer be heard. Fog tends to form when this temperature norm is reversed—when the air is cooler near the ground than it is above. Known as a "temperature inversion," this makes sound waves bend back down toward the ground, rather than up into the atmosphere.

This inversion of the normal temperatures might occur when the low air is cooled during a clear winter's night, as the ground loses

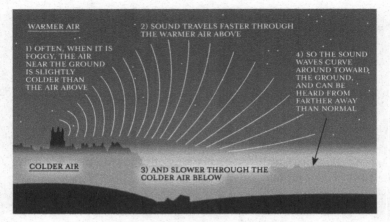

This is why you can the church bells from further away when it is foggy.

the day's heat rapidly into the atmosphere above. Or it might be due to air drifting in over a particularly cold lake or ocean current. Whatever the cause of the fog, the differing air temperatures mean that the sound travels more slowly through the colder air near the ground than it does through the warmer air above. And it is this localized inversion of the temperatures that bends the sound waves downward. With sound hugging the ground, the bells can be heard much farther away.

You might think that the Second Way of the Wave doesn't really apply to sound waves traveling through air since they are not passing through a boundary "from one material to another." But all that is needed for the waves to change direction is for them to change speed. This can as easily happen when the medium they are in varies in a gradual way that affects the wave speed, such as the air temperature changing with elevation. There needn't be any abrupt boundary, nor a different material, so long as the waves' speeds change.

Historically, sailors have taken advantage of sound refraction to lessen the danger of ships getting too close to land as thick fog rolled in. Before the advent of radar and GPS, such conditions could, and often did, spell disaster for mariners in coastal waters. But the increased distance that sound waves traveled along the sea surface due to the temperature inversion did allow sailors to perform a crude sort of echolocation in a bid to avoid being wrecked: they would call out into the fog and listen for the echoes off cliffs. By listening for the direction from which the reflection came, and the number of seconds it took to return, they were able to form a crude picture of where the coast was: the shorter the delay, the nearer the land. The way the cold air near the water refracted the sound back down meant that they could also hear the echoes from farther away than they would normally.

Sailors shouting at the cliffs

I'm not obsessed with aliens and deserts, but I was pleased to learn that an attempt to use the principle of refraction to listen to what was happening on the other side of the world helps to explain the mystery of the Roswell Incident of 1947, which made a sleepy little town in the New Mexico desert the UFO capital of the world.

An understanding of refraction will help put to rest one of the most enduring conspiracy theories of all time: why the debris from a crashed UFO was found near the town and its discovery covered up by a nervous U.S. military. The reason refracting sound waves are connected with UFO crashes can be traced back to one scientist.

UFO spoiler alert

During World War II, Dr. Maurice Ewing, a geophysicist at Woods Hole Oceanographic Institution in Massachusetts, made a discovery into the way sound travels through the depths of the oceans. An expert in the use of sound waves to map the geology of the seabed, Ewing had been asked by the U.S. Navy to research the behavior of sound under water, as this was such a crucial factor in submarine warfare. In 1943, he proved the existence of a "deep sound channel" some 3,000ft below the ocean surface (depending on latitude). This channel traps underwater sound waves within it, ensuring that they travel much farther than they do at other depths. And it works on the principle of refraction.

Within any typical region of mid-latitude ocean, the speed that sound waves travel through the water near the surface is about 3,400mph (the latitude is relevant since water temperature varies greatly between the equator and poles). The dropping temperatures as you descend mean that it slows down to more like 3,330mph at a depth of around 4,000ft. Below this level, the water temperature stops decreasing with depth but the water pressure continues to increase. The effect of this increasing pressure is that sound becomes faster again. By depths of about 16,000ft, the speed of sound is back up to something like 3,440mph. The deep sound channel, at around 3,000ft, is the level at which the sound travels most slowly (it is deeper in the warm waters of the tropics and shallower toward the poles). Due to the effect of refraction, it is also the depth around which much of the sound-wave energy is trapped, unable to spread upward or downward, only horizontally.

Consider a humpback whale calling within the deep sound channel. Normally, sound waves would spread away from it in ever-expanding spheres. But within the sound channel, parts heading toward the surface speed up and bend back downward, while parts heading into the depths also speed and bend back upward. The *Epic whale* "ceiling" of warmer temperatures and "floor" of increasing *songs* pressures mean the waves spread out not in spheres but in ever-widening cylinders. Spread through a smaller volume of water, the humpback's song can be heard by other whales within that channel at great distances—perhaps many hundreds of miles. While marine biologists have speculated since the 1970s that humpbacks and other species, such as northern bottlenose whales, might use the deep sound channel to communicate with each other, it is still a matter of speculation that they actually do.

Whether or not the whales had known about it for millennia, Ewing was the first human to identify the deep sound channel. It became known as the SOFAR, or "sound fixing and ranging," channel and the U.S. Navy enthusiastically funded Ewing's research into its military potential. He proposed a network of underwater microphones, or "hydrophones," that could be used to locate pilots who had crashed in the ocean. The stricken airman would release into the water a hollow metal ball, known as a "SOFAR sphere." This would sink and, upon reaching a depth of around 3,000ft, would implode due to the water pressure, creating an underwater sound wave within the channel. This could then be picked up thousands of miles away, and its location determined by a process of triangulation, comparing the readings from different hydrophones.

As post-war military attention focused on the Soviet Union, Ewing, now at Columbia University, was asked to look into ways of detecting Russian nuclear tests. It had already occurred to him that the same principle should apply to the atmosphere as in the ocean: there ought to be a sound channel in the air, as a result of the way the atmospheric temperatures change with altitude. Might this atmospheric sound channel be a way of hearing booms from the other side of the world?

On average, the air becomes colder with altitude throughout the "troposphere." This is the lowest region of the atmosphere, reaching up to around 7 miles above the poles and more like 11 over the equator. The boundary between the troposphere and the stratosphere above is marked by increased concentrations of ozone and other gases that absorb the sun's heat more readily than those below. So, as you rise through this boundary region, the air stops getting colder before it starts to warm up again in the lower part of the stratosphere. In other words, there is a region of cold air at the top of the troposphere, sandwiched between warmer air below and warmer air above. Since the speed of sound varies only with the air temperature, this region acts as an aerial equivalent of the underwater SOFAR channel. Sound waves within the sandwich of colder air will tend to be trapped since those parts of the waves expanding upward or downward will speed up in the warmer air and so bend back toward the middle. Just like under water, the altitude of slower sound speed traps the waves' energy within it.

Ewing was convinced that this atmospheric sound channel would transmit the sound waves from a Soviet nuclear test right around the world, and he managed to make the chief of staff of the U.S. Air Force feel the same. To monitor nuclear activity, they would just need to listen within the right altitude range.

But since the atmospheric sound channel is in the region of 45,000ft, this was easier said than done. Ewing proposed a system of microphones suspended from large balloons; data from the microphones could be relayed via radio signals either to the ground or to specially equipped planes. This top-secret plan, developed by Ewing with academics from other universities, became known as Project Mogul.

The research and development was carried out in high-security, secluded laboratories and by test launches from remote bases, one of which was the Alamogordo Air Field in the desert of New Mexico, not far from where the first American nuclear test had taken place in 1945. So sensitive was the development of this new surveillance program that the military staff involved at Alamogordo were required to use code words to identify the project, which was overseen by researchers flown in from New York University, and were only informed of their immediate responsibilities. They had little or no idea of how these fitted into the project as a whole. Nor was information about Project Mogul shared with staff at the Roswell Army Air Field, about a hundred miles to the northeast.

There was, however, one aspect that could never be kept secret: the whopping great balloons. These were often joined together *Not very* into enormous trains, almost 700ft long, which sometimes *top secret* contained as many as thirty balloons, along with microphones, radio boxes and silver-foil hexagonal radar targets so that they could be more easily tracked. That they had to be launched during daylight hours to work best also made secrecy rather difficult. To make matters worse, their movement could not be controlled, so the balloons were at the mercy of the winds.

Test balloon launches began at the end of November 1946, continuing into 1947, when Project Mogul was scrapped. It was soon discovered that nuclear blasts on the other side of the world could be detected much more economically, not to mention discreetly, using seismographs. To try to keep the research under wraps, the military took great care to locate and retrieve any crashed balloons and equipment. They even found the UFO reports on the local radio stations helpful in finding them, for silver hexagonal disks floating around in the sky were not going unnoticed by the locals. Sometimes, however, they failed to find the wreckage. Such was the case with one particular test flight on June 4, 1947.

Ten days after its launch, a few miles northwest of Roswell, New Mexico, rancher W. W. (Mac) Brazel and his eight-year-old son,

"We come in search of sound waves."
ABOVE: A 70ft polyethylene balloon of
the type used by Project Mogul over
New Mexico to test equipment for
listening to the Russkies.
RIGHT: Not only did the balloons often
have a flattened shape, appearing
like flying disks when seen from the
ground, they tended to have silver-foil
radar reflectors dangling from them.

Vernon, "came upon a large area of bright wreckage made up of
rubber strips, tinfoil, a rather tough paper, and sticks."[8] At first they
thought little of it, leaving it where it was. But some weeks later,
having heard in the media about the mysterious "flying discs" that
people had reported seeing in the area, Brazel decided to contact the
sheriff in nearby Roswell about the debris. The sheriff immediately
called up the Roswell Army Air Field and an intelligence officer was
sent out with Brazel to collect the unidentified debris. Along with
his colleagues at Roswell, he was unaware of the top-secret balloon
test program being run from the Alamogordo Air Field.

The Public Information Office at Roswell issued a press release to the local papers, stating that "the many rumors regarding the flying saucer became a reality yesterday" as the base's intelligence office had taken possession of mysterious equipment. The following morning, July 8, the *Roswell Daily Record* ran a front-page headline: "RAAF Captures Flying Saucer on Ranch in Roswell Region."

The debris was then sent to the Fort Worth Army Air Field in Texas to be inspected. Whether or not the brigadier general there was aware of Project Mogul is unclear, but he certainly poured cold water on the unwelcome publicity, announcing that this was all a fuss about nothing: the debris discovered by Brazel was simply equipment from a crashed weather balloon. The *Alamogordo Daily News* was invited to take photographs of the wreckage, or at least those parts of it that were consistent with this claim.

Press interest waned once the story turned out to be a non-event. But inconsistencies in witness accounts, combined with rumors within the military that there had been a high-level cover-up, added fuel to the wave of contemporaneous "UFO" sightings, and encouraged conspiracy theorists to claim that wreckage of a flying saucer from outer space had been hushed up at the highest level.

The *Roswell Daily Record*, July 8, 1947: every newspaper editor's dream story.

A full and exhaustive investigation conducted in 1995 by the U.S. Air Force concluded beyond any shadow of a doubt that the debris was just test equipment for a highly classified atmospheric-sound-channel listening device designed to pick up the sound waves refracted around the world from Soviet nuclear tests.[9] The report convinced some of the UFO researchers, but others merely concluded that the U.S. Air Force would say that, wouldn't they?

The Roswell Incident has now become so inextricably woven into UFO folklore that it has become embellished with claims of alien bodies having been recovered and taken for autopsies at the top-secret "Area 51" military base in the Nevada Desert. Every Fourth of July weekend, the city of Roswell holds a UFO Festival, where UFOlogists meet for talks and panel discussions. *Will Smith in a balloon* The incident has become enshrined in popular culture. The plot of the Hollywood sci-fi movie *Independence Day* hinges on the military having managed to rebuild the wreckage recovered at Roswell into an alien spaceship, which is used to defend Earth against an invasion from space. If only the filmmakers had stuck closer to the facts, we would have seen Will Smith confront the enormous alien mother ship, armed with a balloon, a microphone and a tin-foil radar reflector.

~

It is not just sound waves that can refract as they pass through air of differing temperatures. The same happens to light. You can see the effect of light waves bending like this when you look down a long stretch of road in the baking sun. The way the light bends at the boundary between the hot air over the tarmac and the cooler air above makes it seem that the background is being reflected off the road. The swirling eddies as the hot air rises produce a shimmering effect as the light is bent this way and that.

And you can also see the refraction of light in the broken appearance of a spoon sticking out of a glass of water. The abrupt change of direction when the light waves speed up as they emerge from the water distorts the image so that the spoon appears broken, its two halves seemingly at different orientations from each other.

Refraction also explains a feature of ocean waves that has always puzzled me: why do they always arrive lined up with the beach, their forward motion perpendicular to the water's edge? If waves come from storms out at sea, you would think that, every once in a while, they'd roll not up the beach toward the shore, but along the length of it, from one end to the other.

That they don't is due to refraction. The shallower the water becomes, the slower the waves travel, so any wave heading inland at an angle to the shoaling gradient of the sandy bottom will slow down more at the end nearer the beach. This will cause the wave to turn and face the land head-on.

It is interesting to consider how wrong it would feel if waves didn't generally roll up the beach head-on like this. It might be hard to say quite what was wrong, but I feel sure that you would *What if the* know that something was not quite as it should be. Most *waves rolled in* of us, whether or not we are wave watchers, have little *sideways?* appreciation of the phenomenon of refraction. We take the effect for granted. What does it matter to us that waves can change direction as they change speed on passing into a differing medium? That refraction explains why ocean waves generally arrive at the beach head-on may not seem a big deal if you never noticed that waves do so in the first place. But that is precisely what wave watching is all about: noticing the hidden in the everyday.

A wave watcher can, of course, simply enjoy gazing unthinkingly at the surf. I think it is, after all, one of the best forms of meditation there is. But becoming a wave watcher in the broader sense is about finding connections, parallels and similarities among very different types of wave: some that are easily seen, like those at the beach; and others that are invisible to us, like sound. The wave-like nature of our world may be subtle enough for many of us to live entirely oblivious to it, but it is also so fundamental that, once you start noticing it, you begin to see it everywhere.

⌒

This brings us to diffraction, which is the Third (and final) Way of the Wave.

Waves spread around small obstacles as if they weren't there, and spread out in all directions when they emerge from a small opening.

The effect of an obstacle on a wave depends, crucially, on its size compared with the length of the wave. When that object is a lot smaller than the wavelength, the obstacle has a negligible effect on the passage of the wave.

In air, the lengths of sound waves extend from a few inches at the top range of our hearing, to a few feet at the bottom. This means that different types of sound diffract in varying ways around everyday obstacles. Things like trees and fences and parked cars are much larger than the inch-long wavelengths of high-pitched sounds but smaller than the feet-long wavelengths of low-pitched sounds. When you are waiting to cross a busy road, there is a cacophony of pitches of sound, from the high hiss of tires on gravel to the low drone of truck engines. But when there are obstacles—corners and walls, say, between you and the road—you tend to hear just the low, rumbling engines. The higher tones are filtered out.

The same principles govern the optimum positioning of your stereo speakers. The bass speakers, or woofers, can be anywhere in the room—down under a table, perhaps—because the long-wavelength sound waves easily spread around the thick objects and corners to reach your ears. The treble speakers, or tweeters, on the other hand, need to be pointing toward you, with nothing in the way, in order for you clearly to hear their short-wavelength sounds.

It is the diffraction of sound waves, too, that helps you determine the direction from which a sound is coming. This is the way the waves spread around another thick object: your head.

You can use one of two possible methods to locate sounds, depending on whether or not they have large enough wavelengths to diffract easily around your head. High-pitched sounds, with wavelengths considerably shorter than the width of your head, will not easily diffract around it to reach the ear facing away. To locate the direction of these squeakier sounds, your brain compares the sound *intensity* reaching each ear to judge the direction by how much louder it is in one than the other. But low-pitched, long-wavelength sounds easily reach both ears since the size of your head is small by comparison and the waves diffract around it without creating any acoustic shadow. For these deeper sounds your brain compares the minute difference in the *time* the sound takes to reach each ear. The waves will arrive slightly later to the ear away from the source, having traveled a slightly longer path around your head. And when I say slightly, that's not English understatement. I mean no more than 0.6 milliseconds.

The exquisite precision with which our brains can detect the difference in timing and intensity of the sounds is why human beings are extremely good at judging the direction of sounds horizontally.* In fact, for sounds coming generally from in front of us, we are able to tell the difference between sources less than 2° from each other. Although we are not as sensitive at hearing quiet sounds and extreme frequencies as many animals, we are as good, if *A fly with* not better, at *directional* hearing than any other mammal, *excellent* including cats, dogs and bats.[10,11] Until ten years or so *directional hearing* ago, we thought we were better at sound location than *any* other animal, bar the odd species of owl. This illusion was shattered, however, by a pesky little parasitoid fly called *Ormia ochracea*, which lives in the southern United States and northern Mexico. In 2001, researchers at Cornell University found that this tiny yellow insect was able to discriminate sound sources at angles as fine as humans can.[12]

For crickets, the acute directional hearing of this fly has deadly consequences. This is because, at night, the female *Ormia* uses her oh-so-clever hearing to home in on the male cricket's mating call. She lands in the darkness nearby, and scurries the last stretch. Before the poor cricket knows what's up, she has deposited hundreds of her larvae on or near him. One or more of these larvae, which are

....

* But, as any scuba diver will affirm, we are terrible at judging the direction of sounds under water. Sound travels more than four times faster through water than air, so the difference in arrival times at each ear is far less noticeable.

like minuscule black maggots, each less than a millimeter long, proceeds to burrow into the cricket's body. It emerges a week later, having fed and grown within its host, ready to face the world. You'd think that the cricket would be glad to see the back of it. But any relief is short-lived, as it promptly collapses in a heap and dies.

If the *Ormia* fly can resolve sound sources to the same degree as humans, at least it doesn't beat us. Or does it? Our ears are about 6in apart, while the fly's are just 0.5mm. The female *Ormia* fly's ability to locate the cricket is so much more impressive than ours because she is so small. With such a tiny distance between her eardrums, the difference in time for the sound waves to reach one ear compared with the other is in the order of 50 billionths of a second, which makes our 0.6 thousandths look rather ordinary. Oh, what a clever little fly.

c

We may be pretty good at judging the horizontal direction of sounds, but we're awful at judging their elevation—that is, their angle above or below us. This is because, having ears that are symmetrical on either side of our heads, once we turn horizontally in the direction of a sound the distance to each ear is the same for all elevations. There is no difference in the sound waves reaching each ear. This is no big deal, because we essentially operate within two spatial dimensions.

For barn owls, however, life is rather different. Their ability to determine the direction in the vertical plane that a sound is coming from is of the utmost importance. Their eyesight, though twice as light-sensitive as that of humans, is of little help *Wonky-eared* in locating small rodents in the dark when they are scur- *barn owls* rying beneath grass and leaves, or even snow. Instead, the barn owl listens. When it is sitting on a branch, trying to work out exactly where on the ground a telltale mouse rustle is coming from, the owl needs to have a keen sense of vertical as well as horizontal directional hearing. Which is why, in case you've never noticed, barn owls have asymmetrical ears. The opening of their left ears is around a half inch higher than that of their right.

Luckily for barn owls—and the many other owls that share this anatomical trait—their asymmetrical ears are discreetly covered with feathers. Unluckily for mice, voles, and shrews, the difference in elevation means the owl can hear the exact direction of their rustling. Even as the owl swivels its head horizontally in the direction of the rustling in order to eliminate any diffraction and so even up the intensity of the sound waves reaching each ear, the sound waves will still tend to reach each ear slightly out of time with each other.

Until it has also angled its head vertically to point toward the sound, the distance from the mouse to each ear is slightly different. Although there is now no difference in intensity, the owl can look up or down until the timing of the sound waves matches to work out the exact position of its prey. So good is its vertical directional hearing that it can be trained to strike a "sound target" in total darkness, with a margin of error less than 1° in both the horizontal and vertical directions.[13]

As one of the Ways of the Wave, diffraction is not something limited to sound. In fact, its effects can be found in all sorts of waves.

You can see the effect of the diffraction of light when you interrupt a shaft of sunlight coming through the window with your hands to cast the shadow of a rabbit against the wall. We all know that the closer your hand is to the wall, the sharper will be the outline of the little bunny. Contorting your fingers up near the window, some feet away from the wall, will produce nothing more than a bunny-shaped blur. Or is it an owl? No, wait, that's an *Ormia* fly, isn't it?

Whatever it is, the shadow is blurred. This feature of shadows is so commonplace that it barely seems to warrant attention, but *Shadow play* it is a very visible example of diffraction. Your hand is an enormous obstacle compared with the wavelength of visible light, which is in the region of 500 nanometers, or 500 millionths of a millimeter. For this reason your hands most definitely cast a shadow—the light waves certainly don't bend around them as if they weren't there. But they do bend around them a little, since waves always diffract to some degree. The farther your hand is from the window, the more obvious the bending of the light waves becomes.*

Diffraction is also a common feature of radio waves. That is the reason why FM radio stations, which broadcast at wavelengths of around a couple of feet, need to use a network of "repeater-station" transmitters distributed across the country. Without having *Rural radio-* lines of sight into the valleys, the hills and mountains would *wave shadows* cast shadows of poor reception, since they are so much larger than the radio waves. This is not a problem for AM, or long-wave, radio stations that broadcast using wavelengths of around a mile. These need only one transmitter for the whole country, since the long waves diffract more readily around hills into the valleys beyond.

And you can even see the effect of diffraction of waves on the surface of the water. An island, being much larger than the wavelengths, leaves a shadow of calmer water behind (see next page).

A pier post, on the other hand, being much smaller than the length of the waves, leaves no shadow behind it. The waves diffract around it as if it weren't there.

∿

An understanding of how ocean waves follow all three Ways of the Wave was essential knowledge for seafaring navigators of the Micronesian islands in the middle of the Pacific. These expert mariners used to read the swells to hold their direction when canoeing from one island to another. They were, without a doubt, the most impressive wave watchers of all—particularly those from the Marshall Islands, halfway between Hawaii and New Guinea. These remote islands and atolls rise no more than a few feet above sea level, so they are extremely difficult to make out from any distance. The navigators therefore had to rely on the stars for

---

* The blurred edges are not solely due to diffraction. Since the sun is not a point source of light, intermediary tones appear at the edges of a shadow when part of the sun is obscured and part of it not.

their bearings and, when these were not visible, they analyzed the waves.

This unique navigational skill, now largely lost, was passed from one generation to the next with the use of stick charts called *mattang*. Made from strips of coconut-palm fronds tied together into a lattice, these showed how the swells were reflected off the islands and how they changed direction as they refracted and diffracted around them. Small shells attached to the stick charts represented the land. Since the swells tend to come from consistent directions, navigators could learn to judge the direction of an island from as far as 40 miles away by the way it affected the wave paths.[14]

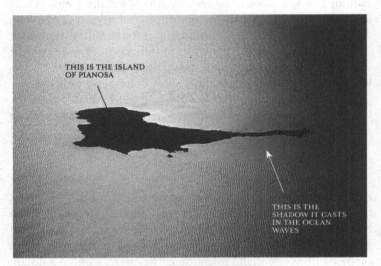

THIS IS THE ISLAND OF PIANOSA

THIS IS THE SHADOW IT CASTS IN THE OCEAN WAVES

PIER POSTS DON'T CAST ANY SHADOWS IN THE WAVES

ABOVE: The island of Pianosa, off the Mediterranean coast of Italy, being much larger than the ocean waves, leaves a distinct shadow.

LEFT: A pier post, being much smaller than the waves, leaves no shadow, for they diffract right around it.

A nineteenth-century *mattang* navigation chart from the Marshall Islands, held in the British Museum. Made of palm fronds, this wave map was used to teach young navigators how ocean swells are reflected, refracted and diffracted as they interact with the islands (represented by the small shells on the left and right).

Actually, it is a little misleading to describe these incredible seafarers as wave *watchers*, since they didn't judge their position by *observing* the waves, more by *feeling* them. "The navigator lies down in the canoe," wrote a missionary in 1862, "pressing his right ear on the floor for several minutes, then he would say to those on board, 'Land is behind us, on one side or before,' and so forth."[15]

The missionary probably got the wrong end of the stick chart with this ear-on-the-floor business. Researchers have since found it to be the rocking of the canoe, rather than any sound, that the navigators were judging: if the stern lifted before the bow, the swell was coming from behind; if port lifted before starboard, it was coming from the left, and so on. In fact, the training of future navigators often involved floating the young man on his back in the sea so that he could get used to the feel of the waves as they rocked him.[16]

No matter how well sound waves demonstrate the Three Ways of the Wave, they might seem to be pretty poor objects of contemplation for a wave watcher. After all, you can't actually see them, can you?

But sound waves had taught me an important wave-watching lesson: sometimes you have to experience waves through their actions, not their appearances. Their waviness isn't always obvious. And, unlikely as it may seem, I felt a sort of empathy with those trainee navigators of the Pacific. The trick, I realized, was not to *look* for the waves but to immerse myself in them.

Which is what I did that May with sound waves. Of course, I was already immersed in them—we all are—but I just paid more attention to the fact. I began to listen to how the sounds around me changed when I entered a room, how my voice and my footsteps *Total immersion* reverberated as their sound waves reflected more readily *in sound waves* off the tiles in the bathroom. At dinner, I stopped to listen to the pure, tuning-fork note from a wineglass as its lip was struck by the handle of a passing knife. I thought about how the delicate glass must have vibrated to produce such a dulcet, silvery note. By contrast, the cymbal-like sound of a pan lid dropped on the floor had no note to it. This was a confusion of vibrations, which had no fixed frequency, and so no clarity of sound—just a background harmonic of the cursing cook.

Although sound waves are generally invisible to us, we can see their effects. If you ever find yourself at a reggae concert, borrow a lighter from a smoker and hold it in front of the bass speaker. You'll see the flame dance in time with the music. It rises, dips and flickers in response to the pressure waves. These are the inherent movements of the sound—the pure physical vibrations, interpreted in an expressive street-dance style by the tiny flame. You can feel the sound waves in your body, too, of course. They make the cavities within your chest vibrate. Here, there can be no doubt about the physical nature of the sound waves sweeping over you.

But if dub reggae is not your thing, if you are more likely to find yourself before an eighty-piece orchestra than a booming sub-woofer, you can still tune in to the physical nature of the music. Close your eyes. Imagine yourself floating on your back in a sea of the sound waves. Feel yourself rock and sway as they pass below you. The tuned skins of the timpani drums have low enough frequencies to stimulate your insides—these are the rolling swell that lifts your whole body to and fro. You can almost detect the separate beats of the deepest of the vibrating drum skins. Not so with the high, trembling vibrato of the string section. This is produced by the musicians rocking their fingers backward and forward as they hold down the string, changing its length ever so slightly, causing the frequency of the note to rise and dip in quick succession. These waves might not shake your innards, but aren't they reminiscent of the way someone's voice wavers when they are consumed by emotion? As you float in the ocean of sound, these waves move you in their own manner, too.

This is not just the physics of blackboards and textbooks. It is the physical root of every feeling that the music stirs within you. Whatever music you are listening to is no more than a succession of sound waves. I don't mean to belittle it, just to remind you of how remarkable it is that such a powerful, multilayered tide of tones and timbers can reach you by such unremarkable means. And to marvel at how all the ambitions that the composer had for the piece, all the emotions the musicians invested in its performance, the intertwining voices of all those instruments, in harmony, in discord, in concert, could possibly be conveyed by just the gentle oscillations of the air within your ears.

The Third Wave

A small river runs along the field near our house. Whenever I am stuck in my work I will wander out to take a look. On one June morning, during a period when we had had no rain for more than two weeks, the river's current was imperceptible. With barely a breath of wind that day, the water's surface was as glassy as a mirror.

When the water is like this I often watch the reflections of the clouds. On that day, these were a drifting succession of fair-weather Cumulus humilis. And although the river's surface appeared at first to be completely still, the mirror image of the clouds revealed its subtle movement. The cloud reflections swayed in an aimless slow dance over the water. But then, as I rounded a slight bend in the river, there was a sudden splash ahead. The reflections upstream started pogoing. What was that? A fish?

By the time I looked, whatever it was had gone. But the little splasher had left a signal: an expanding train of ripples, spreading out across the still water. This broadcast carried a message. These ripples were semicircles, emanating from a point on the opposite bank. They led back to a small hole just above water level. This was no fish. It must have been some sort of water shrew, water vole or—I don't know—water hamster, or something. Whatever it was, it lived in that hole in the bank, and had just given itself away with those ripples.

This got me thinking.

Waves aren't just the movement of energy. They also carry information. This may not sound very romantic, but I suppose it is better than the alternative description of "a disturbance propagating through a medium." The point is that any wave inevitably carries some clues about whatever disturbance got it started. Even though it didn't intend to, our enigmatic river creature communicated information about its movement and its whereabouts both by the sound waves from the little splash of water as it clambered from the river and by the ripples spreading out across the water's surface.

This was when the cloud reflections were still slow dancing.

The cloud reflections had now settled into that side-to-side shuffle you see people do on the dance floor when they're feeling too self-conscious to commit themselves. As I watched, I remembered something that the renowned angler Chris Yates had told me. He said you do a sort of upside-down cloudspotting when you are out fishing because of all the time you spend staring down at reflections of the sky. To the seasoned angler, fish will often unwittingly communicate their presence by producing ripples on the water's surface. So does that make anglers wave watchers as well as cloudspotters? Might it even be possible to identify a type of fish from the particular pattern of surface waves it produces?

∿

"I'm always watching for those surface ripples when it's really calm," said Chris when I called him about wave watching. "It's especially good to look out for them at night with a moon. A fish will move, unseen by you, and because the lake is so still you suddenly see this ripple come by. It's usually one big ripple, with a couple of small 'ripplettes' in front and behind."

Sometimes, he told me, these little undulant messages can be quite subtle, such as when a fish hasn't broken the surface of the water, perhaps when it is feeding on the bottom and it turns to catch a shrimp or something. By kicking its tail as it turns near the surface "it makes that lovely purl, which will send out a ripple."

At other times, a fish will break the surface, often just with its lips as it snatches something floating on the water. "A trout always makes that distinctive circle of ripple," explained Chris. "So *Fish-lip waves* when you're looking along a fifty-yard stretch of stream, you always try and get the light right so you can see the slightest wave opening up on the top, which could be a trout taking a fly."

But the most dramatic of these aquatic broadcasts is when a fish takes a flying leap out of the water, perhaps to clear its gills of mud. "When a big carp has been feeding on the bottom—grubbing around, say, for bloodworm larvae in the silt—its gills can get quite clogged. If you are on a big lake, you might not hear the splash, but you can see the ripple when the surface is still."

One legendary night at Redmire Pool in Herefordshire, back in 1985, Chris caught a 51lb carp. For fifteen years, this was the biggest freshwater fish caught in England. "I was sitting on the dam at one end of Redmire when I heard a fish crash in the shallows, right at the top of the lake, two hundred yards away. It seemed such a long time after the splash that I finally saw the reflection of the moon wobble on the water's surface as the wave arrived."

Chris decided that if the carp jumped again he would time how long the ripples took to reach him. "It did jump twice more and the little wave took two and a half minutes from the fish rolling to the ripple breaking up the reflection of the moon. Two and a half minutes to travel the two hundred yards.

"By the curve of the ripples, you can follow them back across the lake or river to the center of the circle and you've found your fish. When I see a sign like that, I always move around to get nearer. It's a real giveaway. Watching the surface of the water is just like looking at a radar screen."

As traveling disturbances, waves convey information about the event that caused them. While this might be an event like *Big Bang waves* the moment when a carp jumped from the water a few minutes ago, it could equally be an event from farther back in time—say, the creation of the universe 13.7 billion years ago. In fact, the electromagnetic waves produced by the Big Bang are to this day still rippling out across space in the form of "cosmic microwave background radiation."

Just as Chris Yates uses his keen angler's eye to read water ripples for signs of fish, so cosmologists in recent years have used special microwave-sensitive space probes to scan the skies for evidence of the key "cosmic disturbances" of the past. These microwaves have told them a great deal about the origin and composition of the universe—in some cases settling arguments about its basic make-up that had been raging for decades. They are the most important evidence that everything began with a bang, rather than having been forever in a steady state.

The background microwaves are measured and observed using floating observatories. The first of these was the Wilkinson Microwave Anisotropy Probe, which was launched by NASA

Looking for ripples from the biggest disturbance of them all: Planck was launched in May 2009 to measure microwaves originating 380,000 years after the Big Bang.

in 2001, into an orbit 2 million miles from the earth. There it measured the slightest variations in the intensity (or heat) of microwaves coming from different directions in the sky. The subtle inconsistencies in this "cosmic microwave background radiation" across the sky have helped astrophysicists build up a picture of how the universe looked in its infancy, calculating the rate at which it is expanding, its density and its age with a much greater degree of accuracy than has been possible before.

Supposedly, some 1 percent of the static picked up by your TV when it is not tuned to a station is due to this cosmic background radiation.* The rest is largely the electromagnetic noise produced by household appliances and the unfiltered chatter of communication signals produced here on earth—the radio waves and microwaves that are flying around us the whole time. Both the Wilkinson Probe and its more sensitive successor, the Planck observatory, are positioned in orbits that are not only far away from earth, but that position them always in its shadow in order

---

* This TV snow is becoming a thing of the past thanks to the ubiquity of digital television and circuits that black out the screen when there is no signal.

to reduce the interfering electromagnetic racket from our planet as well as the radiation from the sun. Trying to detect cosmic microwaves any nearer to the earth would be like looking for delicate trout-lip ripples on choppy water on a blustery day.

~*~

It is not just fish that betray their presence by the waves they produce. We are all walking disturbances, forever unintentionally broadcasting waves of one form or another and thereby blowing our cover to whoever may be able to tune in.

So whether it is the rustling of a mouse rummaging around in the grass, or the faintest susurration of feathers as an owl makes its swoop, every move of every creature produces acoustic waves of some sort. The visible appearance of any animal is, of course, a disturbance of light waves. Some animals have developed sensitivity to electromagnetic waves of lower frequencies than those that humans can see. While we can feel through our skin the warm infrared light emitted by an animal as body heat, we cannot see it; whereas the jumping pit viper of the Central American rainforests has pits situated between its eyes and nostrils that are sensitive to infrared waves, helping it to strike at its rodent prey with exquisite accuracy.

Such unintended wave communiqués might seem regrettable when an animal ends up as something else's dinner, but where would animals be if they didn't produce waves—indeed, if they *couldn't*? Somewhere at the end of an evolutionary cul-de-sac, I should think, feeling very lonely. For communication of one sort or another is crucial to the survival of every species due to the collaborative nature of reproduction.

Sound waves are clearly a great way to chat up the opposite sex: from the shrill serenade of the male chaffinch in search of a mate at the start of spring to the profound, infrasonic rumbling of the female African elephant, on heat every five years, which is largely inaudible to us but attracts bull elephants from miles around.

The iridescent colors produced as a male peacock's feathers scatter the light waves are not exactly great camouflage, but they don't half impress the ladies. Fireflies flash light-wave signals from their abdomens for much the same reasons, while the cuttlefish produces mesmerizing, shifting colors with the 20 million pigment cells on its skin in order to say "come hither" to mates, "back off" to opponents, and (when used to blend with the background) as little as possible to predators.

But what about humans? Just like any other animal, we attract mates by our capacity to make waves.

Of course, there are the words we say to each other, but even more important for sexual interaction is the tone and timber of our voices. Without even knowing it, a man might modulate his voice to sound deeper and more resonant, to mimic the natural sound qualities of a strong and protecting frame. *Sexy waves* A woman might modulate hers the other way, adopting a fluttery voice to sound as if she's in need of protecting. Or she might adopt a more husky tone to sound sexy. (Such a tweak of timber may be suggestive of the type of broad who drinks, smokes, doesn't hold back, and is up for anything and everything.)

We manipulate light waves for sex, too. Is not the red of lipstick and rouge mimicking the blood to the skin associated with sexual arousal? And why is the color of an iconic Ferrari red? This may seem a crass way of coming onto the opposite sex, but there are clearly many females for whom a Ferrari Testarossa is an aphrodisiac. (I suppose the "redhead" translation of its Italian name *might* refer to the hair of a fiery temptress, as the makers suggest, but I can't help thinking it is suggesting something else.)

Of course, it is not *all* about sex. Every word we say to each other, every melody we hear, every film we watch, every book, newspaper and facial expression we read reaches us by means of sound or light waves. They are the go-betweens—ever-present waves that are constantly seen *and* heard, but rarely noticed. But fundamental to modern-day human communication is one group of waves, of which visible light is just a tiny subset. Our information age is mediated entirely by them. They are the electromagnetic waves.

You already know the color, don't you?

I wish I could tell you that they're called something more friendly. Some people do call them "EM waves," but that's not much better. Given what amazing phenomena they are, these waves are in serious need of a rebrand.

.•ᴄ

Electromagnetic waves range from radio waves, which have the longest wavelengths, through microwaves, infrared waves, visible light waves, ultraviolet waves and X-rays up to gamma rays, which have the shortest wavelengths. The waves in the middle part of *Very long waves* the spectrum—the infrared, visible light and ultraviolet *and very short* ones—come pouring out of the sun. But waves of *all* the *waves* different sizes, from the very longest wavelengths to the very shortest, are given off by the various stars, galaxies, black holes, and hot gases that are strewn across our universe.

No one has yet found any evidence of a limit to the size of electromagnetic waves. The shortest observed gamma rays have wavelengths of around a billionth the size of a molecule, which I have some difficulty picturing. Estimates of the greatest radio wavelengths vary from the distance between the earth and the sun to a thousand times that. If the numbers mean anything to you, that's a range of known wavelengths from 10^{-18}m to 10^{11}m. (These wavelengths, like all the others in this chapter, refer to electromagnetic waves traveling through a vacuum. Passing through anything else, they slow down and bunch up into shorter wavelengths.)

In the middle of this inconceivably broad spectrum of wavelengths is the tiny band that is visible light. From around 700–750 nanometers (thousandths of a thousandth of a millimeter, abbreviated as nm) at the red end to around 400–450nm at the violet end, these all have wavelengths a bit less than a hundredth of the width of a human hair. And they are made of exactly the same stuff as radio waves, X-rays or any of the other types of electromagnetic wave. The only way they differ is in the distance from one electromagnetic peak to the next, and thus in their wavelengths.

A rule of electromagnetic waves is that the shorter the wavelength and the higher the frequency, the more energy the wave contains.* And it is those with larger wavelengths, which contain less energy, that we use for telecommunications. This might seem surprising, for why would waves that carry *less* energy be better for carrying information than the high-energy ones? Surely the more energy a wave contains, the farther it will go and the stronger the signal will be? The answer is that the waves with very short wavelengths are just too dangerous.

Indeed, high-frequency ultraviolet light, X-rays, and gamma rays are so energetic that they can (and regularly do) permanently alter the molecules they encounter by knocking electrons out from the atoms, a process known as "ionization." Prolonged exposure to ultraviolet and X-rays can cause cancerous malfunctions of living cells; and even momentary exposure to the more energetic gamma rays can kill off cells, which is why these are the waves used in radiotherapy to destroy tumors.

It is at the other end of the electromagnetic spectrum that you find the waves upon which our telecommunications age is founded.

Allow me to introduce the waves of electromagnetic communication. The very longest are the radio waves that were once used by the Cold War superpowers to send messages to submarines deep in the oceans. With wavelengths of almost 2,500 miles, they are the only electromagnetic waves able to penetrate more than a few feet of seawater. Being a good conductor of electricity, salt water absorbs all but the very longest radio waves. Producing signals at the extremely low frequencies needed for wavelengths of this size is very expensive and requires enormous transmitters. Synchronized U.S. broadcasting stations in Michigan and Wisconsin each had aerial cables running along telegraph poles, stretching from 14 to almost 30 miles. The sites required their own dedicated power stations. The Soviets, meanwhile, had a single transmitter near Murmansk.

In order to produce radio waves with such huge wavelengths, the equipment on both sides relied on burying poles deep into the ground and using the earth itself as an antenna. With the ending of

---

* This is one way in which electromagnetic waves differ from other waves, whose energies depend on their amplitudes, or heights. It will be described further in "The Eighth Wave."

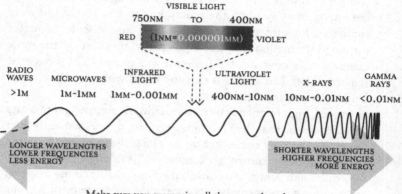

Make sure you memorize all these wavelengths,
as you will be tested on them later.

the Cold War, there was less and less justification for the huge expense of running these transmitters; instead, the subs started coming up near the ocean surface to communicate with HQ. There they can use more practicable higher frequency, radio-wave transmissions, like the rest of us, with wavelengths of between a few miles and a few feet. These are the packhorses of the information age, used to carry signals for radio (obviously), television, baby monitors, garage-door controllers, heart-rate monitors, avalanche-warning beacons, aviation radios, standard time signals, and the little chips that stop people from swiping things from shops, to name but a few.

Next down the scale come the electromagnetic waves of between a few feet and a millimeter in wavelength, which are generally known as microwaves. These don't just heat your food. At much lower intensities, they are used by your mobile phone and for your laptop to communicate with a Wi-Fi network. They also mediate your cell phone's Bluetooth connection, that GPS thing that sends you down dead-end streets in your car, and any long-distance phone calls that need to bounce off satellites. In fact, all communications between Earth and satellites are via microwaves—usually at the shortest end of the band of wavelengths, between 4in and 1mm.

It might come as a surprise that microwaves play such a central role in communication devices if, like me, you've always thought of them as good for little more than heating or defrosting instant food. The ones used in micro-wave ovens, with wavelengths of 12.2cm, heat up the food by causing the water molecules (and, to a lesser extent, fat compounds) to rotate backward and forward. Having positive and negatively charged ends, these H_2O molecules will swivel back and forth like maniacal compass needles, trying to stay aligned to the electric field that shifts backward and forward 2,450,000,000 times a second as the microwaves pass into the food. The glass plate and ceramic bowl aren't heated by the waves, as they don't contain water molecules.

But many other wavelengths of microwaves have frequencies that are either too high or too low for much of their energy to be absorbed by water—be it water in a tub of instant food or suspended within the atmosphere. And since they don't interact with the charged particles of the "ionosphere" like many other sizes of radio waves, they pass readily through our atmosphere. Not only are microwaves therefore a natural choice for communicating with satellites, but they are also the preferred means of communi-cation with spacecraft far beyond Earth's orbit. It is via microwaves that we stay in contact with the most distant man-made object in the universe: NASA's deep-space probe *Voyager 1*.

Launched on September 5, 1977, *Voyager* is now more than 10 billion miles from Earth, officially outside our solar system, and moving farther away by a million miles a day. Like all electromagnetic waves, the microwave communications between Earth and deep space travel at the speed of light. The time it takes for information from *Voyager 1* to reach us is currently almost fifteen hours, which puts the delay you sometimes hear on long-distance calls into perspective.

VOYAGER 1

10.22
BILLION
MILES

THE EARTH
THE SUN

"Can you hear me?
It's not a great line."
Man's most distant
communication ever
uses microwaves.

But echoing conversations are largely a thing of the past now that telephone conversations, like internet traffic and cable TV, are transmitted by infrared waves passing down optical fibers. With frequencies higher than microwaves, and wavelengths of between 1mm and 750nm (millionths of a millimeter), these electromagnetic waves are easily blocked by many types of solid material. This explains why infrared is also the electromagnetic wave of choice for TV remote controls. Were these to employ waves that passed through walls, arguments between neighbors would become a daily—if not hourly—occurrence.

Remember that all these different types of electromagnetic waves vary only in size and frequency. The amazing thing is that they are all the same waves, just at different scales. And though we don't ever think about it, they are all around us, passing through us everywhere we go: incessant messages, signals and information overlapping, intersecting, combining and passing on their ways. We actually see just a tiny slice of the whole electromagnetic cacophony, and yet it is all there. In the words of the bongo-playing physicist Richard Feynman:

> So all these things are going through the room at the same time, which everybody knows, but you've got to stop and think about it to really get the pleasure . . . the complexity, the inconceivable nature of Nature.[1]

Which makes you wonder how the hell a baby monitor manages to tease out the one wave signal it needs to pick up from the electromagnetic mess. If all these overlapping waves are just different-sized versions of the same thing—the electromagnetic equivalent of everyone and everything shouting at once on an unimaginable scale—how can a device possibly pick up the one essential signal: that of your baby crying?

All these electromagnetic communication devices depend on one particular phenomenon: resonance. This is the way that waves, be they electromagnetic or any other, interact with the world around them. Born of the intimate relationship between waves and vibrations, resonance is one of the most pleasing wave phenomena of them all.

～

A struggling musician, Marv, has just bought a new guitar.

He's not all that good at the guitar, which is probably why he is struggling, but he manages to make a kind of living busking in the town center. After several months of "The House of the Rising Sun," he's finally earned the coinage required to upgrade his old instrument. With great pride he brings the new one home to his apartment and tunes it up for the first time.

As Marv plucks the guitar's D-string, it vibrates back and forth at some 147 times a second. This vibration leads to a corresponding one over the surface of the shiny, lacquered soundboard that forms the front panel of the instrument. This is what generates the majority of a guitar's sound. The waves of the D note radiate outward from the vibrating soundboard. They spread through the dingy room, filling the chilly space with a warm, bright note.

The relationship between waves and vibrations is so intimate that you can barely separate the two. Vibrations cause waves, but the relationship is not just one way. Waves produce vibrations in return—particularly periodic waves, which are, after all, *Waves* moving patterns of vibrations. And this is the case with *and vibes* the warm D note emanating from the soundboard of Marv's new guitar. Over in the corner behind his back sits Marv's old guitar. It is scratched, beaten-up and covered in stickers. It has accompanied him on every jam session for the past five years, earning the coins to buy its own replacement. It may have been cast aside in place of the newer model, but the old strummer's still in tune.

Its own D-string also naturally vibrates at 147 times a second.* So when Marv plays the note and sends 147-hertz sound waves through the room, they cause the D-string on the old guitar to vibrate gently in sympathy. The frequency of the waves matches the rate at which the old string naturally vibrates. The succession of arriving compressions and rarefactions of the arriving sound wave

---

* In fact, the sound waves will also make the A- and the G-strings ring out a little, since they share harmonic frequencies with the D-string. But this effect will be much less pronounced than the resonance of the D-string.

chimes with the string's natural movement. Each adds to the effect of the last, building up the movement in the string until it starts to sing out its own note. And now the guitars are playing the same note together.

This is much like Marv pushing his son on the swings in the park. "Higher, higher!" shrieks his son. Marv's been on his feet all day and he's got a few things on his mind, so he isn't really giving it his all. He could pull the swing up high, as high as his shoulders, and then let go, but he doesn't. Instead, he starts with a gentle push and then builds up, pushing in time with the swing's natural rhythm to gradually increase the arc. Marv times his pushes subconsciously. Were he to push any faster or slower than the rhythm of the swing, than its "natural frequency" of swing, he would be colliding with his son one moment, pushing thin air the next. The swing would never build up and his son would complain that he was a bad dad.

And so it is with the old guitar across the room. The other strings don't ring out like the D-string does under the influence of the note he just played. The fundamental rates at which they vibrate, their natural frequencies, differ from that of the D-string. Each crest of the sound wave is to them a mistimed push on the swing, which doesn't add to its vibrations in any cumulative manner.*

....

* In fact, for both guitars, this is just the string's main, or "fundamental," frequency. At the same time, it also vibrates at other "harmonic" frequencies. These produce higher notes whose frequencies are the fundamental multiplied by two, three, four . . . and so on. Each is the sound of a different type, or "mode," of vibration of the string. When you pluck a guitar string, not only does it vibrate the fundamental mode like this:

which, for the D-string, is the sound of the note D. At the same time, overlaid on this vibration, are ones that look more like these:

These simultaneous vibrations produce the second and third harmonics, which are the notes D and A within the octave above the fundamental. Since these and many other modes of vibration occur at once on a plucked string, there are always other harmonic notes that sound at the same time as the main one, adding to the warmth of the tone.

But each string of the old guitar does have one frequency at which it resonates most vigorously: the frequency that matches its fundamental mode of vibration. So as Marv begins to play a tune on his new instrument—still "The House of the Rising Sun," I'm afraid—the strings of his old one gently and naturally react to certain notes. It's as if the old guitar has played the tune so many times that it has learned to accompany Marv by itself. Of course, he doesn't notice this sympathetic backing track. He can't hear how the old strings resonate beneath the sound of those he is strumming, but resonate they do, subtly ringing out as their frequencies are played. Only when he finally places his hand against the strings to silence his new guitar does he hear the reverberant notes of the old.

~~

This phenomenon of one string affecting another by means of resonance has been harnessed in the design of a number of musical instruments. One example is the Baroque stringed instrument called the viola d'amore.

While the design varied somewhat from one instance to another, the vast majority of viola d'amores made in the eighteenth century, at the height of the instrument's popularity, shared a distinctive feature: below the six or seven strings, which were bowed and fingered like any violin or viola, ran a corresponding set of "sympathetic strings" underneath the fingerboard. These were tuned to the same pitches as the upper ones, but were untouched by the musician. They would vibrate in sympathy with the sound waves produced by the played strings. Their sympathetic resonance gave the instrument a distinctively warm and rich tone—one that Leopold Mozart, Wolfgang Amadeus's father, claimed "sounds especially charming in the stillness of the evening."[2]

The sound holes on the front of the viola d'amore were in the shape of flames or swords of Islam, which might suggest that the instrument had roots in the Middle East (its name may, in fact, derive from the "viola of the Moors"). Rather than having a scroll like a violin, the part beyond the tuning pegs was usually decorated with an elaborately carved head. This was typically that of

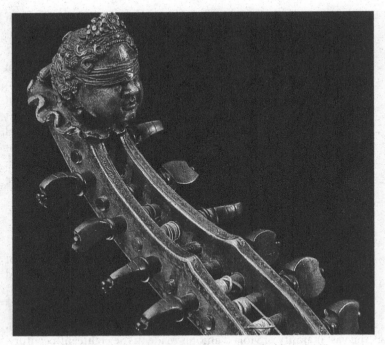

"Wait, don't tell me. I know this one . . . It's Vivaldi, isn't it?"
A viola d'amore made by Jean Baptiste Deshayes Salomon, *c.* 1740.

a blindfolded Cupid—an appropriate symbol for an instrument whose strings each have hidden partners with which they vibrate in sympathy and, at times, in unison.

In the nineteenth century, resonance became a symbol not only of love but also of human empathy, thus of one person's sensitivity to another, particularly on an emotional or intuitive level. Take the Calvinist essayist Thomas Carlyle's 1828 description of his fellow Scotsman, the poet Robert Burns: "Tears lie in him, and consuming fire; as lightning lurks in the drops of the summer cloud. He has a resonance in his bosom for every note of human feeling."[3]

And the metaphor of resonance was sometimes used to suggest communication in an unspoken, almost telepathic, manner when two people are in tune with each other. In the opening chapter of Emily Brontë's 1847 classic *Wuthering Heights*, for instance, Mr.

Lockwood intuitively understands his landlord, the surly Heathcliff: "Some people might suspect him of a degree of under-bred pride; I have a sympathetic chord within that tells me it is nothing of the sort: I know, by instinct, his reserve springs from an aversion to showy displays of feeling."[4] The same could be said of Jack Kerouac's *On the Road*, written more than a hundred years later, when he described the relationship between William Burroughs and his wife, Jane Vollmer, as having its own peculiar resonance: "Something curiously unsympathetic and cold between them was really a form of humor by which they communicated their own set of subtle vibrations."[5]

Human resonance

And by the late 1960s and 1970s the metaphor came full circle. Resonance was once again about being on the same wavelength as your surroundings—no longer crystal spheres with planets attached, but more likely swirly colors of oil projected on the walls of a nightclub. The Beach Boys sang about picking up good vibrations, while Tom Wolfe, in *The Electric Kool-Aid Acid Test*, claimed "something's getting up tight, there's bad vibrations."[6] In 1966, Timothy Leary called for people to "Turn on, tune in, drop out," later explaining that "'Tune in' meant interact harmoniously with the world around you."[7]

~c~

This year, watching a band play in my local village hall, I found myself thinking about resonance throughout the performance. This was not because they were playing eighteenth-century viola d'amores—on the contrary, one of them was performing on "hang drums," which are a little like steel pan drums in the shape of flying saucers, and were invented in Switzerland in the year 2000. The reason I was thinking about resonance was simply because of the effect that the music was having on me.

Perhaps I just don't get out enough, but the music was so good that I felt as if I really was vibrating inside in sympathy with it. Indeed, I felt a mild euphoria. Did it have the same effect on the other members of the audience? The band was well received, but I'm sure the feelings they elicited were as numerous as the people in the crowd. This seems to fit with the metaphor of resonance—that, just like differently tuned strings, what gets me going might not resonate with you and, indeed, on another occasion, might not have a big effect on me either. Quite why music affects us so profoundly is as much of a mystery as it has always been.

The ancient Greeks were fascinated by the apparently magical power of music, particularly with its ability to soothe or excite the human soul. Pythagoras's interest in both math and music led him to discover that there is a mathematical basis behind musical harmony. He found that the most harmonious musical intervals—an octave, a fourth, and a fifth—are between notes with the simplest mathematical relations. On a stringed instrument, for instance, two strings of the same weight and tension will sound harmonious together when their lengths have mathematically simple ratios. Strings an octave apart (separated by twelve semitones) will have lengths in the ratio of 1:2—one string will be twice as long as the other—while for a perfect fifth (seven semitones up), the lengths will be in the ratio of 2:3, and for a perfect fourth (five semitones), 3:4. The discovery that there is a straightforward mathematical basis for what sounds beautiful to us was so inspiring that it led Pythagoras and his followers to propose that the mathematics of musical harmony might actually explain life, the universe, and everything.

He took the discovery of the mathematics of musical harmony and ran with it—rather a long way, if truth be told. Pythagoras argued that the mathematical relationships between the movements of the moon, planets, and stars across the night sky are those of musical harmony, for they produce a music of the spheres—vibrations or notes that we can't hear, but that are in perfect harmony with each other.[8] Moreover, we humans produce continual, inaudible music, and are instruments capable of sympathetic vibrations. According to Pythagorean theory, this explained why normal, audible music could affect man so profoundly. The human instrument could also, on occasion, tune in and reverberate with the inaudible music of the spheres. Which is presumably what the Beach Boys were singing about.

~c~

Resonance might serve as a rich metaphor for communication but it plays a far more workaday role in enabling the world of telecommunications.

A "resonant circuit" lies at the heart of every mobile phone, Wi-Fi laptop card, car radio, or a baby monitor. Such a circuit has a natural frequency at which it oscillates most readily, which *Electromagnetic* is determined by the configuration of resistors and other *resonance* electronic gubbins. These oscillations are the movement of electric current backward and forward through the wires of the circuit. And, just like someone on a playground swing, the components mean that there is a certain rate at which the current most easily swings to and fro through the wires.

Radio waves are what do the pushing. Being electromagnetic, they encourage the electrons in the metal of the aerial to move up and down as they pass over it; they "induce" a tiny current in it. This is something that radio waves do as they pass over any piece of metal—the poker by the fireplace, the springs of your bed as much as an aerial—but the movement of electrons, or currents, is absolutely minuscule. Unless, that is, the aerial is connected to a resonant circuit and the to-and-fro pushing on the electrons by the radio waves matches the rate at which the circuit resonates, its "resonant frequency." Just as the gentlest pushes on a playground swing, perfectly timed to match its natural rate of swinging, will soon build up the movement, so the feeblest push on electrons by the radio waves at the circuit's resonant frequency will build up a detectable current.

The unfathomable mess of electromagnetic waves passing over the device's aerial will all tend to encourage the electrons within the circuit to oscillate back and forth. But only those electromagnetic waves that match the circuit's resonant frequency will chime with it. These are the only well-timed pushes on the swing.

In this way, the circuit responds selectively, filtering out from the incredible cacophony of electromagnetic waves just that signal it is designed to pick up. The radio wave that matches the resonant frequency is called the "carrier wave." This is what the circuit responds to, the frequency that builds up to a far greater current than all the noise from all the other waves. Overlaid onto this carrier wave (often rather like ocean wavelets on top of a large ocean swell) is the all-important signal that is being communicated: the radio music, the telephone conversation, or the baby crying. Because of the way that waves add together, the receiver just needs to take away the carrier wave to be left with the signal (like flattening out the broad sweep of the ocean swell to be left with the wavelets).

For some devices, such as a baby monitor, the resonant frequency is fixed; for others, such as a radio, turning the tuning dial changes the frequency it resonates at. Likewise, turning the tuning key of a guitar changes the natural frequency at which a string vibrates, and so the sound waves to which it will sympathetically respond.

Resonance is essential to the age of telecommunications because it is the means by which a particular signal can be teased out of the electromagnetic melee.

I had a regular experience of the phenomenon of resonance when I used to work from a shed at the end of our tiny garden in northwest London. My shed was octagonal, had black felt tiles on the roof

Garden-shed resonance

and was really cozy when it was raining. In my opinion, the sound of rain is wholly underappreciated. I found the raindrops drumming on the roof produced a pleasing white noise that invariably helped to focus my thoughts. The same could not be said of the sound of passing helicopters.

This part of London has more than its fair share of street crime and I would often hear the beat of chopper blades as police surveillance teams pursued knife-wielding thirteen-year-olds through the backstreets. When one flew over, the sound would often resonate in the shed. Being a hollow chamber with an open window, the shed behaved a little like a huge musical instrument. Just as the air column in the tube of a clarinet is stimulated by changing oscillations in air pressure from the vibrating reed that you blow against, so the air in my shed was stimulated by the pressure changes of the helicopter beating sound. A hollow chamber with openings has a resonant frequency just as a guitar string does. (This is the frequency you hear when you blow across the open end of a glass bottle.) And it turned out that the shape and materials of my shed gave it a natural frequency that happened to coincide with that of the pulsing chopper blades. They were at the resonant frequency of the shed. Unless I had the window shut, the beating could become louder inside than out. Sometimes the beating sound was enough to make my ears pop, which served as an uncomfortable reminder that sound waves comprise changes in air pressure.

Perhaps I should have written to my local politician to complain that I was the victim of resonant sound waves resulting from the youth crime wave sweeping the neighborhood. I should have demanded that something be done about this crime wave before it did my ears in. Either that or I should have written a why-oh-why letter to the shed's Swedish manufacturers, accusing them of resonance negligence, for surely they must have helicopters in Sweden?

~c~

I doubt that anyone ever made such a complaint to the tomb and burial-chamber builders of the Neolithic era. Toward the end of the Stone Age, around 4000–2000 BC in Europe, this was the period when most of the Megalithic monuments still standing today were built. Besides the earthworks, standing stones and stone circles, such as Stonehenge, many took the form of chambers covered with earth or stones, accessed by one or more passages, often *Burial-chamber* in a cruciform shape. Though these were often found to *resonance* contain bones, it is by no means clear that their main purpose was as burial tombs, for they might have served more as temples, perhaps related to the worshipping of the spirits of ancestors. The emerging field of "acoustic archaeology," which studies the acoustic properties of ancient structures, suggests that the people who designed and built these chambers took great care over the resonant properties of their underground spaces.

In an effort to shed some light on the possible function of these ancient structures, Paul Devereux and Robert G. Jahn, respectively an author and a professor at Princeton University, studied the acoustic properties of a number of prehistoric chambered structures across England and Ireland.[9] They investigated Chûn Quoit in Cornwall, a single-chambered stone structure that was covered in earth, Wayland's Smithy in Berkshire, a stone-chambered long barrow, as well as Newgrange, a large passage tomb with a cross-shaped chamber, and two of the cairns in the Loughcrew Hills, all within County Meath in Ireland. At each location, within these enclosed chambers, they played tones using loudspeakers, adjusting the pitch to find the frequency at which the intensity of sound vibrations built up most within the space, and at which the sound grew louder than at other frequencies. The resonating of the chambers was caused by the sound waves traveling down the passages, bouncing off the ends, and adding up as they overlapped themselves on the way back. When they compared the frequencies that caused the greatest reverberation like this, the researchers discovered something rather surprising.

Even though the structures varied greatly in size, shape and construction materials, they all resonated within a very narrow band of frequencies: 95–112 hertz. This falls comfortably within the human vocal range, at least that of a male baritone. Human remains found within structures like these have led to an archaeological consensus that they served as burial chambers. Might their very specific and shared resonant qualities indicate that, as well as or before being used for burial, they were originally used for some sort of ritualistic chanting, speculated the researchers? Chanting or singing at the resonant frequency of the chambers would have the effect of enhancing the volume and reverberation of the voice and might "create a commanding sense of the presence of supernatural agencies, whether gods or ancestral spirits."[10]

Researchers at Reading University studying the acoustic qualities of Camster Round, a Neolithic passage grave in Scotland, analyzed an exact scale model of the cairn to investigate its resonant properties. They found that, since it is shaped as a narrow passage leading into a round chamber, the structure ought to resonate rather like the air in a bottle when you blow across the neck. Known as "Helmholtz resonance," this is when the air in the bottle expands and contracts as a whole, to produce the tone. The scale model led the researchers to suppose that the Neolithic structure in Scotland

My local Neolithic burial chamber, Stoney Littleton Long Barrow, in Somerset. Might it have behaved as a resonating chamber that gave the sound of chanting that extra oomph?

ought to behave like an enormous bottle. They found that the chamber could be made to resonate in the same way as a bottle by producing sound within the chamber (which seems rather more likely than Stone Age worshippers desperately blowing across the chamber entrance). The model indicated that the Helmholtz resonance of the structure would be in the region of 4–5 hertz.[11] But wait a minute. That is far below the human vocal range and, for that matter, the tonal range of musical instruments. Humans can't even hear tones below 20 hertz. Surely this destroys any theory about Stone Age worshippers making sounds resonate in Camster Round while performing their rituals?

The researchers didn't think so. There might be a way to build up sound vibrations even at frequencies too low to hear as a note. A pure tone is made up of pressure pulses that don't individually register as sound to our ears. Only when they follow each other fast enough to vibrate our eardrums more than twenty times a second do we hear them as a pitch. But if you bang a drum four or five times a second, you are producing audible sounds that repeat at a frequency of 4–5 hertz. Each bang is a sound-wave pulse (like the pulses that make up a pure tone), but is followed by the reverberation of the drum's skin—something we certainly can hear. We can hear the rapid beats of someone drumming four times a second, even if they aren't rapid enough for our brains to join them together into a note with pitch.

It was time to take a drum to Scotland.

The researchers assembled an audience in the burial chamber with them and whacked a drum four times a second (a rate of 4 hertz). The audience reported feeling unusual sensations during the bouts of drumming. In particular, they felt as if their pulses and breathing patterns were being influenced by the sound. Some said that, were the drumming to go on for too long, they thought they might end up hyperventilating. Fewer complaints were reported during equally loud sequences in which the beats were at slower rates—too slow to make the space resonate.

Such reports are clearly subjective, but when NASA researched the effects of vibrations on the body that needed to be considered in rocket design, they found that different parts of the adult frame also resonate at certain frequencies. When vibrated at these rates, the various internal organs vibrate with greater intensity, causing intense "decrements in performance and discomfort."[12] And the resonant frequency for the human torso? Four to five hertz, the very same as the one the researchers identified for Camster Round.

Might Neolithic worshippers have felt that they were communicating with spirits, gods or ancestors on account of drumming causing infrasound resonance? Research during the 1970s found that frequencies in the 4–5 hertz region not only make people feel dizzy and uncomfortable (on account of the vibrations building up and jiggling around their innards), but also make them feel drowsy, with sensations of falling and swaying.[13] Might the Stone Age architects have designed the structures specifically to exhibit these resonant qualities? Might the resonating sound have seemed otherworldly, even induced infrasonic vibrations within their own bodies that altered their consciousness? And might the neighbors have told them to keep the racket down?

~e~

Have you ever woken in the middle of the night with the horrifying realization that you haven't the faintest idea what an electromagnetic wave is? Me neither. But since these waves are everywhere around us, I felt that it might be a good idea to find out.

The main quality that sets them apart from "mechanical waves," such as sound and water surface waves, is that they don't depend on a physical medium through which to travel.

This means you might need to adjust your idea of a wave. After all, it is not easy to conceive of a wave without there being some medium for it to pass through—without something that is physically *moving* as the wave passes by, most obviously when you consider water waves. To talk of an ocean wave without any water for it to pass through is clearly nonsensical. This is a type of mechanical wave because it is a traveling pattern of physical movements within a medium—in this case water.

The dependence of sound waves on a medium is not so obvious, but can be demonstrated. The pressure waves that are sound can only exist when there is something physical for them to *Sound waves* travel through—be that something air, water or the wall *in a vacuum* between you and your death metal-loving neighbor. If you don't believe me, suspend a little bell in a sealed glass jar, suck out all the air with a vacuum pump (I'm sure you have one on hand), and give it a shake. You'll hear (or rather *won't* hear) the proof.

For, surrounded by a vacuum, the bell's vibrations cannot reach you. This is why "In Space No One Can Hear You Scream," or, for that matter, hear your spaceship blow up. The dramatic deep-space explosions of sci-fi movies should really take place in complete and utter silence, as there aren't enough gases present to spread waves of compression and expansion. Filmmakers might like to consider sustaining the excitement of their galactic battle scenes by reintroducing Charlie Chaplin–style piano accompaniments.

So much for the music of the spheres.

Electromagnetic waves, by contrast, travel happily through the vacuum of space. The sun may not be heard, but it is most definitely seen and felt. Were you to replace the bell in that vacuum jar of yours with a white-hot filament, you'd have no difficulty seeing the light waves coming off it. (You would, after all, have made an early incandescent lightbulb that contained a partial vacuum before such bulbs came to be filled with inert gases.)

So, if electromagnetic waves can travel without the need for anything physical to pass through, what exactly is doing the waving? Physicists will tell you that it is a combination of electric and magnetic fields that are oscillating. But, to be completely honest with you, I find it hard to picture what they actually mean.

Physicists and mathematicians can describe and predict the behavior of electromagnetic waves with exquisite precision. They know all about their formation, the way they propagate from one place to another, how they interact with matter and electric, magnetic, and gravitational fields. They know these waves intimately and absolutely from a mathematical point of view, thanks

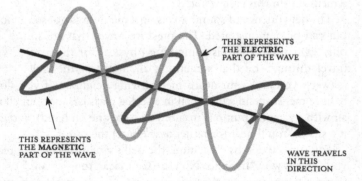

Electromagnetic waves go this way and that way at the same time.

to the brilliant Scottish mathematician and physicist James Clerk Maxwell, who in 1864 derived a set of equations that defined these waves in terms of oscillating electric and magnetic fields. So well do Maxwell's equations describe the behavior of electromagnetic waves that they are still in use today.

Thanks to Maxwell, the mathematical description of electromagnetic waves is now straightforward. The tricky bit is understanding what these oscillating electric and magnetic fields actually *are*. An electric field is some property of space that exerts a force on an electric charge; likewise, a magnetic field on a magnet. They are intimately linked, like two sides of coin, since the movement of an electric charge produces a magnetic field, and the movement of a magnet produces an electric field. And this interdependence seems to be how they pass through a vacuum.

It is not easy to understand quite what electromagnetic waves are. So take a deep breath: we're going in . . .

A moving pulse of electric field causes a corresponding moving pulse of magnetic field (orientated at right angles to it), which gives rise to a moving pulse of electric field again (still at right angles), and so on. Since the changes in one field give rise to the other and vice versa, the two make their way across the void like a pair of silver-screen characters on one of those levered railroad handcars. The energy moves through space as a transverse wave composed of two components—one electric, the other magnetic—oriented at right angles to each other.*

Traveling through a vacuum like this, the speed of all electromagnetic waves is the same: the speed of light, which is around 186,000 miles per second or 300,000,000 m/s.** They are the fastest things we know of, and are generally considered to define the cosmic speed limit, which nothing can ever exceed.

⌒

* In some situations, electromagnetic radiation behaves not like waves at all, but like moving particles that don't have any mass, known as "photons." This other way of looking at the electromagnetic spectrum is discussed in "The Eighth Wave." In this chapter, they're waves, and nothing but.

** It's more like 299,792,458 m/s if you want to get fussy about it.

In 1981, the controversial biologist Dr. Rupert Sheldrake proposed a form of resonance very different from that of vibrating strings, oscillating circuits or resounding burial chambers. He called it "morphic resonance,"[14] and suggested that it might explain, among other things, how the complexity of organisms develops from embryos. Sheldrake proposed that, once a pattern—be it biological or physical—has occurred somewhere and at some time in nature, there is a higher chance of it recurring. Self-organizing systems, such as cells, crystals, organisms, and societies, draw on some sort of self-perpetuating, collective "memories," he claimed, which influence and inform how future occurrences of the systems are organized.

As one of the supports for his hypothesis, Sheldrake cited a study, from 1954, and conducted over a period of twenty years, into the learning abilities of fifty generations of lab rats.[15] The time taken for the rats to learn a task was compared from one generation *Rats get smarter* to the next in an effort to see if skills were passed on in any way. The study found that successive generations of rats descended from those who'd been taught a task each learned to perform the task more quickly than their ancestors. You might infer that the rats genetically inherited some sort of improved ability to pick up the task from their trained ancestors.

But you'd be wrong. The surprising part of the study was that it found a parallel increase in the speed of learning within control generations of rats—ones descended from ancestors who had never learned the task. Each subsequent generation of rats seemed to

learn the trick faster than previous ones, even when they were the first in their genetic line to have been taught it. It seems that you can teach an old rat new tricks, but you'll have a lot more success if you teach a young rat old tricks.

Sheldrake suggested that the same principle affects how easily people learn tasks too. Not only do you get better at a computer game the more often you play it, but the easier it will become for others to

"Are you on my wavelength?"

learn it, too. Don't let your teenage children hear about this, as it will give them a new excuse for spending hours a day in front of their X-Box: they're doing it for the benefit of teenage gamers in future generations.

Sheldrake hypothesized that some sort of resonance might explain why successive generations get better at learning things—whether rats in the lab, or teenagers on the sofa. He made no attempt to explain how such a morphic resonance might work, merely posited it as a way of accounting for this phenomenon, and many others to do with complex systems. Just as the natural, physical vibration of something can stimulate a corresponding resonant vibration in something similar by means of waves traveling between the two, perhaps physical and biological systems influence each other's structures in analogous ways. After all, points out Sheldrake, "Atoms, molecules, crystals, organelles, cells, tissues, organs, and organisms are all made up of parts in ceaseless oscillation, and all have their own characteristic patterns of vibration and internal rhythms."[16]

The idea does sound rather like it came out of the academic wing of the acid generation. No sooner do you wonder whether some sort of cosmic vibrations might mediate this morphic resonance than you start sounding like Timothy Leary. Who knows if Sheldrake is onto something? But ever since he first proposed the hypothesis, he had the scientific old guard up in arms.

John Maddox, the editor of the respected science publication *Nature*, famously condemned Sheldrake's hypothesis of morphic resonance. He wrote that Sheldrake's book, *A New Science of Life: The Hypothesis of Formative Causation*, was "the best candidate for burning for many years,"[17] and in a 1994 BBC television interview he claimed that Sheldrake was "putting forward magic instead of science."

Maddox added that Sheldrake's book should be condemned "in exactly the language that the popes used to condemn Galileo, and for the same reasons: it is heresy." (The fact that Galileo turned out to be right, and the popes wrong, seemed curiously to have been overlooked.)

Maddox and other skeptics of Sheldrake's hypothesis argued that it was so broad in its scope and claims as to be impossible to disprove—that it was effectively unfalsifiable—placing it beyond the realms of science and into those of pseudoscience.

But it wasn't always scientists in search of falsifiability who attacked Sheldrake. In April 2008, Sheldrake sustained minor inju-*Guinea pig* ries when he was stabbed in the leg by a member of the *gets even* audience at a talk he was giving in Santa Fe, New Mexico. A Japanese man was arrested and charged. He later claimed that he believed he had been a "guinea pig" in Sheldrake's mind-control experiments using "remote mental telepathy."[18]

~c~

The sound artists Bruce Odland and Sam Auinger employed the principle of acoustic resonance in a work of art called *Harmonic Bridge*, which has been installed since 1998 on the Highway 2 overpass outside the Massachusetts Museum of Contemporary Art. A pair of aluminium "tuning tubes" are attached to the railings of the overpass, right by the traffic. At specific positions within these tubes, microphones pick up the sound waves, which are amplified to a loudspeaker on the pedestrian walkway below.

The tubes act as huge musical instruments, much like didgeridoos built by plumbers, which respond to the vroom, screech, and drone of the traffic as it slows and accelerates at the traffic lights on the overpass. The tubes, open at each end, were made 16ft long so that the fundamental frequency at which they resonate is close to the 33 hertz of a very low C note, three octaves below the middle C on a piano.

Just like the vibrations of a guitar string, other frequencies resonate in the tubes as well, these being the higher harmonics of the C note. The artists experimented with the positions of the microphones in each tube so that, along with the fundamental C, they emphasize different harmonics. Odland and Auinger adjusted them until they liked the sound coming from the loudspeakers. In one tube, they put the microphone one-sixth of the way along, a position where it picks up more of the 6th and 12th harmonics—both notes of G, one above and one below a middle C. In the other tube, they placed the microphone two-sevenths of the way down,

Harmonic Bridge, near MASSMoCA, North Adams, Massachusetts, by O+A (Bruce Odland and Sam Auinger). TOP: A "tuning tube" on the overpass. BOTTOM: A speaker beneath the bridge plays the note at which it resonates.

emphasizing the 7th harmonic, a B-flat. Being just a semitone away from C, this is perhaps the least harmonious of the harmonics. The artists liked the way it added a bluesy feel to the sound below.

The two crude musical instruments purify the unpleasant, dirty mess of noise produced by the traffic, extracting only those pure frequencies that resonate in the air columns. The sound emerging from the speakers below is a mix of each, forming a harmonious blend of notes—predominantly C, G, and B-flat—the strength of the different harmonics waxing and waning with the changing drone of the traffic. Deeper road noise, like the engines of buses and trucks, causes more of the lower notes to resonate in the tubes, while higher sounds, like those from cars, motorcycles, and the voices of passing pedestrians, excite more of the higher harmonics. To quote the artists, the sound has "altered the emotional landscape under the bridge in a humane way, and has reclaimed a forgotten urban space as inhabitable."[19]

The phenomenon of resonance results from the intimate relationship between waves and vibrations. You can sum it up with three simple facts:

> *Waves and vibrations are intimately related. Vibrations cause waves and waves are traveling patterns of vibration.*

> *Whatever vibrates tends to have characteristic natural frequencies at which it does so most readily.*

> *Waves with frequencies that match the natural frequency with which something vibrates will tend to cause its vibrations to build up and grow more and more pronounced.*

Tap a good-quality wineglass to hear the pure note it produces, which is the sound of its natural mode of vibration. Sing that note (in any octave) from over the other side of the room. Come back and listen up close, and you will hear the glass still ringing in sympathy with the sound waves you produced.

Resonance may be nothing more unusual than the child of a wave and a vibration, but it always seems to feel rather mysterious. It can feel as if tiny wave-like miracles are taking place every time it occurs. Sing a succession of different notes and the glass will selectively respond only to the ones that chime with its natural frequency.

And the phenomenon doesn't just occur between physical vibrations and mechanical waves. Resonance also occurs between oscillating electric currents and electromagnetic waves, such as radio waves. The resonant circuits within telecommunications receivers—whether baby monitors, or spacecraft communication networks—are the means by which we are able to tease out the unfathomable jumble of signals carried by the electromagnetic waves around us.

Every wave carries news: information of some sort about whatever created it. This is part of what a wave is, since each and every one is caused by something. It is set off by some disturbance, vibration, oscillation—by *something*—be it momentary or ongoing.

Occasionally, when there are few other waves around to interfere, it is easy to notice the wave and read its news. A fish jumping at the other side of the lake on a still, clear night makes the moon's reflection dance across the water, and tells the angler where to cast his line. But most of the time there's a cacophony of other waves muddled up with the ones we are interested in, especially in the case of modern telecommunications. The very term "broadcast" is derived from the action of scattering seeds, throwing *What a* them in broad swathes to cover as much of the ground, *blooming racket* and as evenly, as possible. The same applies to many of the broadcasts of electromagnetic waves so that they can be picked up over a large area. At every moment of every day, we walk through a soup of electromagnetic messages sweeping as waves through the space around us: radio stations, emergency-service communications, international time signals, mobile phone conversations, Wi-Fi transmissions, satellite link-ups, air-traffic control, text messages, speed-camera motion detectors, TV channels, weather radar . . . It is mind-boggling to contemplate them interfering and interacting with each other—combining here, dividing there—as they overlap and interweave.

From this cacophony, we pluck the one seed of information we want by means of resonance. It is how we are able to tease out the messages and signals that waves inherently carry. It mediates the effortless extraction of order, clarity, and sometimes, beauty from the tremendous racket of life.

The Fourth Wave

WHICH GOES WITH THE FLOW

The Eisbach, or "ice brook," is a man-made canal that feeds off the River Isar in the German city of Munich. It travels down a tunnel that runs beneath the city-center streets, before entering the Englischer Garten park, where it emerges from the tunnel mouth in a gushing torrent.

A split second after the moss-green water rushes out from the arches below the street, it hits an abrupt ridge in the concrete base of the channel. This kicks the flow upward to form a "standing wave," 3ft or so high, where the current rises and dips in a smooth arc, before tumbling in upon itself in a confusion of white water. On long summer afternoons, a gaggle of tourists can be found on the pavement of the Prinzregentenstrasse above, peering over the stone balustrade, snapping photographs and pointing.

They're not watching the wave so much as the surfers riding it, as they steer their surfboards along the stationary wave from one side of the icy current to the other. Locals have been surfing the wave on the Eisbach since the early 1970s, right there in the middle of the city, just on the other side of the traffic lights. Since the channel is only about 35ft wide, there is really only room for one surfer at a time, so they line up to wait their turn. Jumping onto the front slope of the standing wave, they ride back and forth from one side of the watercourse to the other. The water rushes beneath them but they, like the wave, don't move downstream with it. Always in wetsuits, because the water is so cold, they perform jumps and spins and spray cascades of icy water up onto the banks.

This form of surfing doesn't involve going anywhere, which means that the spectators can see every move from close by—something that *Surfers going* is never the case when surfers are riding ocean waves. Only *nowhere* when wave riders lose balance, or catch the current with the edge of the board, are they swept downstream in the surging torrent, at which point the next surfer jumps on to take over.

Standing-wave surfing is a growing sport in Munich. It is also performed on another branch of the same River Isar, the Flosskanal, at a place called Flosslände, in the south of the city. This wave is not quite as challenging as the Eisbach, since the water doesn't flow as fast and the wave doesn't rise as high. But it is there, where the channel is wider and there's more room for spectators, that they hold the Munich Surf Open each year, on the last Saturday in July. It is the only surf competition in the world to take place 200 miles from the nearest coast—thanks to standing waves that form within a flowing current.

～

So, what is a standing wave?

The answer is a wave that doesn't travel from one place to another. Normal, "progressive" waves start at some source and spread as vibrations through a medium, which itself needn't be displaced overall. The crests and troughs of a standing wave, by contrast, remain fixed in place. This seems quite an unwave-like thing to do. So why do standing waves stay put? It can be for one of two, quite

Surf culture, Munich-style.

different, reasons, depending on whether the medium they form in is itself flowing, like the current in the Munich watercourse.

A standing wave that *doesn't* rely on a current is the sort you find in musical instruments. It is the means by which it produces a pure note. Blow over the mouthpiece of a flute and normal progressive sound waves will travel up and down the column of air within.* At each end of the flute, the sound waves are reflected back the other way. This means that identical sound waves travel backward and forward through the same column of air—those bouncing off the far end passing through those coming from the mouth-piece end. The compression (increased pressure) or rarefaction (decreased pressure) of the air at any point along the length of

....

* In the case of a flute, the current of air that you blow doesn't flow down through the flute, but *across* the opening at the mouthpiece. It causes a fluttering effect—a form of resonance—of rapidly rising and falling air pressure within the air column near the mouthpiece. These changes in pressure travel as sound waves along the air column within the instrument.

the instrument is the result of the sound waves traveling along from the mouthpiece and back from the far end adding together, or "interfering." Where two regions of maximum compression, the "peaks" of the two sound waves, overlap, the air becomes doubly compressed. Where the compression from one wave overlaps with the rarefaction of another, a "peak" crosses a "trough," the two cancel each other out and the air is at normal pressure. And the result of this interference of identical sound waves traveling up and down through the same column of air? A stationary pattern of "nodes" where the two waves always cancel each other out to produce minimal fluctuations in the air pressure, separated by "antinodes" where their compressions and rarefactions add up to produce maximal fluttering of the air pressure.

The combination of identical progressive waves moving in opposite directions like this sets up a standing-wave pattern of vibrations, determined by the length of the air column. This pattern doesn't move along the length of the flute, but remains fixed in place. It is the means by which the instrument resonates at a pure frequency, the very mechanism that causes it to hold a consistent note, as normal progressive sound waves spread out from the pressure fluctuations at the openings. Lifting fingers off the holes forces the antinodes of maximum pressure variation to form at different points along its length, and so changes the resulting tone.

All this might sound a rather dry description of the deliciously warm and sonorous tone from a well-played flute. Will you now be unable to listen to the opening of the third act of *Carmen*, surely one of the most beautiful examples of a solo flute in opera, *Spoiling* without worrying about nodes and antinodes? Have I *the classics* ruined it for you? I could explain that standing waves also form in stringed instruments, when transverse waves travel up and down the length of the strings, bouncing off the fixed ends and interfering with each other. But I don't want to ruin Bach's cello suites for you, too. This sort of standing wave doesn't have a name to distinguish it from the other sort, but it might be fitting to call it an "interfering standing wave."

A more visible example can develop on the surface of the water in bays, estuaries, and harbors that are open to the sea at one end.

Waves entering from the sea can reflect off the land at the coastal end, travel back in the opposite direction and, in so doing, interfere with new waves as they arrive. Most of the time this just results in a confusion of peaks and troughs on the water of the harbor, but occasionally, when the wave periods (and so their speeds) are just right, the reflected crests and troughs will meet with the arriving ones at the same points in the harbor, causing a fixed pattern of standing waves to develop. Known as "seiches," these stationary patterns of rising and falling water consist of nodes, which are where the incoming and reflected waves always cancel each other out to leave a fairly flat surface, separated by antinodes, where they add to make the surface rise and dip substantially.

At certain wave periods, dependent on the dimensions of the harbor, quite dramatic seiches can build up, with the rise and fall of water dashing moored boats against harbor walls and, in extreme cases, throwing them onto the shore. Violent seiches can *Harbors and* also be formed when earthquakes disturb lakes and other *soup bowls* such enclosed bodies of water. The water rises and falls as the waves bouncing from one shore to the other overlap. (The same can happen to the soup in your bowl if you stumble and make it slosh from side to side as you carry it to the table—rising and falling antinodes form at the sides of the bowl and a node of fixed soup level remains in the middle.)

But it is not an interfering standing wave that the Munich surfers ride. Theirs is the second type of standing wave: one that forms within a current. It also lacks a distinguishing name, so I call it a "fluent standing wave." You can see a miniature version whenever you watch the water in a stream glide over a rock that lies just below the surface.

In the lee of the obstruction (just downstream of it), there is either a rough jump upward in the level of the water or, when the current is less swift, a smooth undulation that stands fixed in position. The obstacle or rock deflects the current upward as it passes over, lifting it in a peak above the equilibrium level before it is pulled back downward by gravity, thereby dipping beneath the equilibrium level to form a trough. Eventually, it settles back to the mid-level farther downstream.

No doubt, during the twilight hours, Mole and Ratty paddle out on strips of bark and ride a fluent standing wave in the river when no one's looking.

～

Fluent standing waves also form within the currents of the atmosphere, where they are sometimes revealed by the appearance of "lenticularis" clouds. These can appear when the wind is deflected upward by an obstacle—a very large obstacle, such as a hill or a mountain—and their name comes from the Latin for "lentil."

This is not as ridiculous as it might sound. The clouds can look rather like individual lentils—even if they are white ones—floating in the sky and often several miles across.

The geological obstacle that kicks the current of air upward is the equivalent of the rock in the stream, or the ridge in the concrete floor of the River Eisbach. When the atmosphere is what meteorologists describe as "stable," the airstream rises and dips in the lee of the peak like the surface of the flowing water. So long as the wind remains steady, the invisible standing wave of air remains stationary downwind of the peak. And when the air has the right temperature and humidity, a lentil-shaped cloud forms at one or more of the wave's crests.

Droplets materialize when the air rising at the front of the wave expands and cools enough for its moisture then to condense. These are what we see as cloud, hovering in position within the stiff breeze, a standing wave of air rendered visible. Having formed at the front of the wave, the droplets shoot along in the wind, only to evaporate away again where the air dips and warms again at the back. The cloud shape appears stationary, even though the droplets are racing through it in the rising and falling air current.

These lenticularis clouds are one reason why mountainous regions make such great places to go gliding. The pilots ride the standing *Surfers* waves downwind of the peaks like airborne equivalents of *in the sky* the Munich surfers, maneuvering their gliders from side to side within the rising part of the air current. The lenticularis clouds act as beacons at the crests of the waves, letting them know

LENTICULARIS CLOUDS REVEAL THE CRESTS OF THE STANDING WAVES

MOIST AIR-STREAM RISES TO PASS OVER MOUNTAIN PEAK

"STABLE" AIR TAKES ON WAVY PATH IN LEE OF MOUNTAIN

You can't see the river of air that forms a fluent standing wave in the lee of a mountain peak, only the lentil-shaped cloud that appears at its crest.

where the invisible river of air ascends, providing precious lift, and where it sinks, its downdrafts threatening to dash their aircraft to the ground.

So fluent standing waves rely on a current to form, while normal progressive waves can easily travel through a stationary medium. But, alas, it is not quite as clear-cut as that. Few aspects of waves ever are. Take ocean waves, for instance. Currents occur throughout all the oceans of the world, so what effect does the flow of water have on normal progressive ocean waves? In some places, it can turn them into monsters.

The Agulhas Current runs in a southwesterly direction down the east coast of Africa. Once it rounds Cape Agulhas, at the continent's southernmost tip, the current encounters waves coming in the other direction that have been generated by storms in the South Atlantic. Upon reaching the oncoming current, these waves are slowed down. You might think that this would mean they become less formidable. You'd be wrong, however, and it was with good reason that Bartholomew Diaz, the Portuguese mariner who navigated this coastline in 1488, named it *Cabo das Tormentas*–the Cape of Storms.

To understand why ocean waves grow in height when they encounter a current flowing in the opposite direction, it might be helpful to catch up with the group of aliens that crash-landed in the previous chapter.

By this stage, having followed the road that they stumbled across, they have entered a small town and found themselves in a shopping mall. The shrieks and screams of the locals have panicked them. Running through the mall as fast as their suckers will take *The aliens* them, they head for the escalators to seek refuge on the *are back* first floor. But, unfamiliar with such contraptions, they make the mistake of running up the down escalator. As the first alien in the line starts running up it, his progress is slowed since the treads are moving in the opposite direction. The panicking alien behind him catches up a little before he jumps on. The same applies to each of the others, and so they bunch up in a tighter group as they struggle up the counter-current of the escalator. Only when alighting from the escalator near the lingerie shop on the first floor do the exhausted aliens spread out again, since each steps onto solid ground a moment before the next.

Something similar happens when ocean waves encounter a current flowing against them. The waves progress more slowly through the opposing current, and so, upon meeting it, the wave crests become bunched together, which is to say that their wavelength reduces. Since there is nowhere else for the water to go, this concertina effect makes the waves grow taller.

Such a concertina of waves is one reason why the swells out at sea off the Cape of Storms can grow so enormous. But there are two other contributing factors. One is the effect of the "Roaring Forties," the band of fierce and relentless winds that circumnavigate the globe to the south, unimpeded by any obstructing landmasses at these latitudes within the Southern Ocean. Blowing against the flow of the Agulhas Current, these winds keep adding energy to

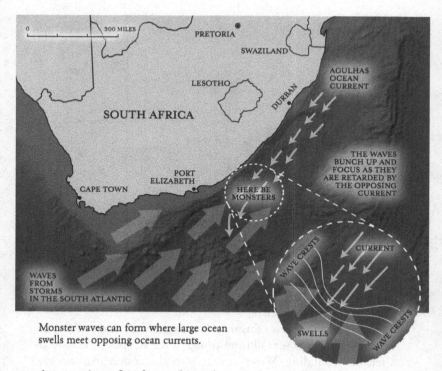

Monster waves can form where large ocean swells meet opposing ocean currents.

the water's surface by pushing the crests onward. The other is the effect the current can have in focusing the wave energy.

Up on the first floor, meanwhile, the shopping-mall aliens have started to hold hands again for solidarity as they run screaming past the shops. The one in the middle of the line is unfit and starts to lag. As he slows, those running full pelt at the ends are pulled in toward the middle.

The same happens with the waves on either side of the Agulhas Current. Like the fitter aliens at the ends of the line, they keep traveling faster, their progress unhindered by any ocean current flowing against them. They are also pulled inward, not because the middle ones are unfit, but because they slow down in the counter-current. This is a form of refraction, caused by the waves changing speed upon moving into a current. The effect is to focus the waves in toward the middle. There, within the current, their

"Oh, *that* scratch? I'm pretty sure it was there when we picked it up." The *Wilstar*, a Norwegian tanker, was hit by a rogue wave in stormy conditions within the Agulhas Current in 1974. Its bows were torn off like the lid of a sardine can.

energy is concentrated into a smaller area of water, causing the peaks to rise still farther.

This is why the waves forming out at sea off the Cape of Storms can grow almost 100ft tall from trough to crest, about the same as a ten-story building. What's more, with the concertina of wavelengths, their faces can become extremely steep. However, such monsters are not the norm. They're described as "rogue waves," and rear up more than twice as high as the prevailing waves.

It seems that they result when two or more storm waves, already tall and steep on account of the current, pass through the same patch of water at exactly the same moment and momentarily combine to form a wave that overshadows the others. Sometimes a correspondingly deep trough precedes the crest. This can be particularly dangerous to shipping, as it is hidden from view until the last moment, when the ship is teetering at the crest of the previous wave.[1] No wonder the seabed near the southern tip of Africa is littered with shipwrecks.

The other day I came across these verses, written in the 1920s, by the American poet Robert Frost:

Sea waves are green and wet,
But up from where they die
Rise others vaster yet,
And those are brown and dry.

They are the sea made land
To come at the fisher town,
And bury in solid sand
The men she could not drown.[2]

This poem set me wondering about sand waves. Are they actually a sort of wave, or do they just look like waves?

Take the little ridges of sand that you feel below your feet as you step into the water at a shallow beach. I've always loved the sensation of these compacted, serried sand ripples massaging my

Serried ridges of wet sand, formed by the ebb and flow of tides up a beach. Do they bear any relation to ripples on the water, other than looking the same?

soles just beneath the water's surface. They are formed by the action on the sand of flowing water as it advances and retreats with each wave rolling up the beach, or when the tide rushes over level sand. The distance between these sandy ripples might be just a few inches. But undulations of over 3ft can form in the sand below a swifter, more consistent current—perhaps where the tide is constricted as it passes between land masses, forming a swiftly flowing channel. When the current is faster like this, and also when the grains of sand are coarser, the ripples can grow to well over 3ft between crests. Anywhere between 3 and 50ft, and these oversized undulations are known by the pleasing name of "megaripples."

"Sand waves" is the name given to even larger undulations. In the strong currents that sweep across the sea floor off the coast of the Netherlands, for instance, sand waves cover the seabed for an area of over 5,500 square miles, sometimes with wavelengths of over half a mile and heights of as much as 60ft. Apparently, these waves migrate along in the current at a rate of between 35 and 500ft a year.[3]

Many of these "bed forms," as they are known, do *look* like sandy waves. They form in currents of water, and they are even called sand ripples and waves. But does that mean that they actually *are* waves?

～

What about the dry relations of these underwater sand ripples— the ones that form when the sand gathers in undulations as it is blown along in the wind?

Just like their wet cousins, these form in a range of sizes. At the smaller end of the spectrum are the endless bands of ripples that appear along the dry, windblown surface of a beach. I watched them up close on a windy day at the beach in mid-July. It was a hot afternoon, but one better suited for a beach walk than sunbathing since the wind carried a blanket of sand grains a few inches above the surface that stung your cheek and filled your ears if you lay down. I suffered the stinging long enough to watch how this barely visible blanket of grain-laden wind, a few inches thick, was making

Are the sand seas of the Sahara covered in a form of slow, dry waves?

the ripples inch along the surface. They even seemed to move like slow-motion water ripples. Were they a form of wind-blown wave? They formed in a current of air like the standing-wave lenticularis clouds, but the sand ripples only stand still when the wind drops and stops blowing them along. A lenticularis cloud, on the other hand, just disappears when the airstream that forms it dies down. So was Robert Frost right? Were these a sort of desiccated, slow-moving wave?

Of course, Frost was talking about sand *dunes*—and not those that have been immobilized, their surfaces now bound together by grass. He was referring to dunes of loose, dry sand that are free to blow in the wind.

These are the dunes you find in areas too arid to support sand-stabilizing vegetation, such as in the endless "sand seas" in Libya and Egypt. In these enormous deserts, the sand is often completely free to be blown about. The result is endless vistas of dunes, which creep along with the prevailing winds, foot by foot, year after year.

Some contours of these sand seas rise in smooth, graceful arcs, like calm ocean swells, while others are peaked and covered in smaller ripples, resembling restless, choppy squalls.

Shifting desert dunes seem as timeless as ocean waves, "giving a hint of that which changes not."[4] But, in growing regions of the world, they also represent time running out. The very modern-day problem of desertification, caused by the effects of overgrazing, deforestation and excessive use of water resources, has led to farmland, roads, railways and villages being overrun by advancing sand dunes.

The nation most threatened by "dune encroachment" is China, where the vast deserts in the north and northwest of the republic are spreading eastward. An Asian Development Bank assessment of Gansu Province in 2001 estimated that dunes had already overtaken 500 square miles of farmland, and that four thousand villages were under threat of being engulfed.[5]

The dunes are coming—run for your lives!

The ancient town of Dunhuang, once an important staging post on the Silk Road trading route that runs through the region, has 1,600ft-high dunes looming at its southern edge. These stand like an advancing mountain range, and surround the dwindling oasis where the Crescent Moon Lake, which was 35ft deep in the 1960s, is now barely 3ft. In the late 1990s, across China as a whole, these dunes have consumed over 4,000 square miles of arable land and grazing pasture each year.[6]

This relentless advance of sand is also a major problem for many other countries, too. Reports within the last ten years have described how more than a hundred villages have been buried in Afghanistan's Sistan Basin, while over the border in Iran the dunes have claimed at least another 124. And desertification is similarly a growing problem in parts of Brazil, India, Mexico, Kenya, Nigeria, and Yemen.[7]

China's battle against the advancing waves of sand, now less than 150 miles from Beijing, has led the government to initiate a seventy-year project of tree planting in an attempt to stem the flow. Dubbed the "Green Wall of China," this will be a 2,800-mile "shelterbelt" of trees. The idea is that the roots of the trees will consolidate the terrain and, with a bit of luck, halt the advance by cutting off the sand supply.

These undulations of sand may be far from any ocean, but they do seem rather like "the sea made land" coming "to bury in solid sand the men she could not drown."

With Robert Frost's words ringing in my ears, the time had come to sort out this whole sand-dune confusion once and for all. The only way to find out whether or not sand ripples, megaripples, sand waves and sand dunes were actually a form of waves would be to speak to an expert.

But where do you find a sand-dune expert? Just ask around until you find someone who goes by the name of an "aeolian geomorphologist."

Dr. Andreas Baas, senior lecturer in Physical Geography at King's College, London, is one such person. It was a little embarrassing how excited I was to get him on the phone. Here, finally, was someone who could confirm that the moving undulations of sand found in currents or water and wind are actually waves, albeit slow-moving, granular ones. Imagine my disappointment, then, when he told me I'd got it all wrong.

"We geomorphologists are always very reluctant to consider dunes to be waves," he told me from the outset, "because the whole mechanism of their formation is very different from water waves."

Baas then proceeded to pummel my dune theory into the sand. He did it in the nicest possible way. But my attempts to argue that these sandy undulations are a form of wave looked feeble from the outset.

We had only just started when he delivered his first blow. Dunes are fundamentally different from waves because the wind doesn't actually lift and distort the surface of the sand in the way that it causes the water to be deformed. On the sea, the water sinks back down to where it started, and the disturbance travels along the surface. In the case of a dune, the wind carries a flow of sand, which starts to collect around an obstacle or slight bump and builds up as a cluster. I realized he had a point when I stopped to consider this difference in how dunes and waves form:

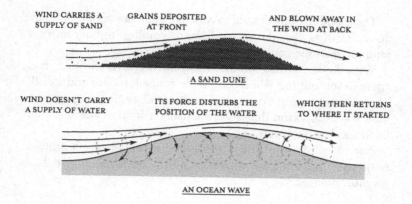

WIND CARRIES A   GRAINS DEPOSITED   AND BLOWN AWAY IN
SUPPLY OF SAND   AT FRONT   THE WIND AT BACK

A SAND DUNE

WIND DOESN'T CARRY   ITS FORCE DISTURBS THE   WHICH THEN RETURNS
A SUPPLY OF WATER   POSITION OF THE WATER   TO WHERE IT STARTED

AN OCEAN WAVE

A dune grows only because it is fed by a current of grains carried in the wind. But an ocean wave does not depend on a current of water.

Ouch. Even so, I came back with a one-two. What about a standing wave that forms in a current of water? Doesn't that form in a flow—and, for that matter, around an obstacle?

"But there you are talking about water flowing *through* the shape of the wave," Baas replied patiently. The sand certainly doesn't flow with the wind through the shape of the dune like a fluent standing wave of water. At any moment, the grains in the middle of the huge mound of sand aren't going anywhere as the wind blows over it. The dune shape is not formed by an overall current of sand, like a sort of horizontal landslide. Only the top layer is lifted and blown along in the wind.

I must admit, I took some encouragement from his admission that it's a little more complicated when you compare dunes to standing waves in currents.

The second reason dunes are fundamentally different from water waves, he continued, is their shape. "Dunes are asymmetrical, while waves have a symmetric shape—the front and the back of the wave are essentially the same."

It turns out that the windward slope of a dune is fairly gentle, around 11°, while the back is a steep slipface of 30° or more. Down this face, the sand periodically avalanches as the dune inches forward in the wind.

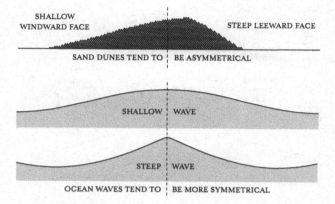

Sand and ocean waves often don't even have the same shape.

This time, I was quick on my feet and employed a masterful defensive move: what about a wave that is breaking on the shore? Doesn't it then have a gentle slope at the back and an abrupt cascading front?

"Yes, I suppose I can see that a little bit more," he pondered. "In a breaking wave, you have an overall flow of water. The water inside the wave is, at that point, actually moving forward with the wave."

Was that a point to me, then?

He soon wiped the smile off my face with his third argument, though: "Of course, unlike a water wave, the actual migration of the dune form is by the transport of material over the top of the surface." In other words, the way the sand moves as a dune edges forward in the wind is fundamentally different from the way the water moves as a wave propagates across the surface of the ocean. The wind deposits sand on one side of a dune and then lifts the grains up the slope and over the crest in a succession of jumps known as "saltation." Periodic avalanches of sand down the back face of the dune every now and then mean that the whole thing advances like a very slow-moving tank track (see next page).

My defense was beginning to crumble. Baas's argument left my dune-wave analogy reeling. You can appreciate how fundamentally different they are when you consider what happens when the wind

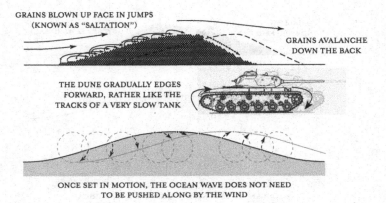

GRAINS BLOWN UP FACE IN JUMPS
(KNOWN AS "SALTATION")

GRAINS AVALANCHE
DOWN THE BACK

THE DUNE GRADUALLY EDGES
FORWARD, RATHER LIKE THE
TRACKS OF A VERY SLOW TANK

ONCE SET IN MOTION, THE OCEAN WAVE DOES NOT NEED
TO BE PUSHED ALONG BY THE WIND

Sand and ocean waves move in completely different ways. Damn.

stops blowing. In the case of the dune, the mound of sand just sits there dead in its tracks. Energy needs to be constantly applied to the system for the dune to move forward. By contrast, when you start a wave on the surface of some water by slapping your hand down on it, the disturbance travels along on the energy you initially gave it. There is no need to keep pushing it along. Likewise, the energy that storm winds transferred to the ocean surface keeps the waves going long after the storm has dissipated. They roll on as ocean swells after the winds have stopped, often for hundreds of miles.

Damn. I couldn't argue with this. I was defeated.

Then I had brain wave. I'd been reading about traffic jams and heard that they travel along the road as sort of waves—not ones that go up and down, of course, but compression waves, traveling patterns of more densely packed cars that move along the road. Since a dune moves along by being fed with sand at one end and depleted of sand at the other, might it be considered like a traffic jam, which is fed with cars at one end and depleted of them at the other?

"Yes, very much so, actually," came Baas's reply. "I guess it all depends on how loose is your definition of a wave," he added, magnanimous in victory.

I was no longer quite sure what my definition of a wave was, but I was sure that it was loose. Very loose.

c

Since leaving my old life in London for the rural pleasures of Somerset, I've spent more time than I'd like on that arterial road of the West Country, the A303. The traffic is bad enough at any time, but if I'm foolish enough to make the journey near the weekend, this artery is in danger of thrombosis.

The road alternates infuriatingly between dual and single lanes. When traffic is heavy, the merging of two lanes into one invariably causes vehicles to back up for miles.

So far, so explicable. But sometimes jams seem to develop from nowhere (and not because there's always some *bastard* who goes barreling up the inside lane as far as he possibly can before forcing his way in at the front).

One minute we're all moving (and getting) along fine—flowing steadily, even if we are rather densely packed. And then, just as I am enjoying the spectacle of clouds over Salisbury Plain, we suddenly slow to a halt and edge along at a walking pace.

What's going on? Is it construction? A breakdown? Some dozy cloudspotter losing control of the wheel? There are no clues, but I feel sure I'll find out around the next bend. And then, without warning, the car in front pulls away and we're flowing again.

What was all that about?

c

Professor Yuki Sugiyama, of Nagoya University in Japan, runs the Mathematical Society of Traffic Flow. Before you rush out to join this society, I must break the news that it is open only to "physicists, engineers, mathematicians, and biologists, involved in the mechanics of traffic flow."

When I called him to ask about traffic waves, Sugiyama explained that there's no need for a bottleneck in order for a jam to develop. "We always have a fluctuation in our driving [so] once the average density of cars exceeds a critical value, anyone can act as the trigger that sets off a jam." In other words, if there are too many vehicles on the highway, the flow of traffic becomes unstable. This means that any typical driver can unwittingly start a jam and, sooner or later, one will.

The reason Sugiyama is so confident that a jam will form on a crowded road in the absence of any obstruction is because he demonstrated it with cars driving around and around (and around and around) a completely obstruction-free circular track.[8]

Like other scientists studying the behavior of traffic, Sugiyama had for many years simulated jams using computer models. But his was the first practical experiment showing them forming spontaneously.

Twenty-two drivers spaced out around the 750ft track were instructed to proceed as steadily as possible, at a speed of 19mph, keeping a safe distance from the car in front.

It took only a lap or two before fluctuations began to appear in the distances between the cars. This wasn't the fault of any driver in particular. Nor was it about bad driving. Everyone's speed varies a little as they drive. But since the cars were packed quite densely on the track, these random fluctuations set off disturbances in the regularity of the flow, which soon built up.

Noticing the car in front to be a little too close, a driver would brake a little, but overcompensate slightly, and slow down slightly too much. This meant that the driver behind would end up over-compensating a little more. In this way, the "disturbance" in the traffic flow built up. In no time, a "stop-and-go wave" developed: a mini-jam, averaging five cars, in which the cars had to stop momentarily before they could move on.

Once they'd worked their way to the front of the jam, the cars could pick up speed to the requisite 19mph once more. At the same time, new cars joined it at the back. So although the cars only ever moved forward, the position of the mini-jam traveled backward through the traffic flow.

This might have been just a modest five-car jam, but it was a jam nevertheless, and it proved that no bottleneck is required to trigger a snarl-up if the cars are packed close enough. In the real world, on real highways, there is a magic number of cars per mile, a universal density, where the flow becomes unstable. This is no different from one country to the next or from one speed

If you think your commute is tedious, you ought to have a go on this road.

limit to another. Measurements of the flow of vehicles on different highways in Germany and Japan have shown that the change from free-flowing traffic to congested traffic always happens when the density of cars reaches forty vehicles per mile.[9] If a motorway becomes anymore crowded, the flow becomes unstable and the inevitable little fluctuations in drivers' speeds soon develop into stop-and-go waves.

The speed at which Sugiyama's mini-jam traveled backward through the traffic flow closely matched the speed stop-and-go waves progress along real highways. Video films from aircraft have shown that, regardless of how large the jam is, the speed a spontaneous cluster of cars propagates backward along the road is always around 12mph.[10] Consider a jam stretching for a couple of miles along the interstate. Assuming that the overall number of vehicles on the highway doesn't change, the position of the jam will creep backward along the road as cars pull away from the front, and others join at the back. An hour later, the 2-mile-long jam will have moved 12 miles or so down the road (or *up* the road, really—in the opposite direction of the cars themselves).

The natural speed of a traffic wave

If it seems counterintuitive that the traffic wave should always move at the same speed regardless of the size of the jam and the overall speed limit of the road, bear in mind that the rate at which the wave progresses depends primarily on drivers' reaction times. It is determined by how fast they pull away when the road in front becomes clear, something that is about the same everywhere, whatever of the speed limit.

Finally, it was time to ask the $64 million question.

Is a wave in traffic a *real* wave? And if so, what sort of wave is it?

"A stop-and-go wave is a cluster solution of a non-energy-conserved dissipative system," Sugiyama explained.

Right. I'm glad we've got that straight.

Eventually, I was able to glean that a stop-and-go traffic wave does indeed move in a similar way to a sand dune. But both are quite different from a water wave—even one that develops in a river or ocean current and, like them, is forming within a flow.

Traffic jams and dunes are "dissipative systems," in that their energy is not enclosed but instead leaks out. Because of this, you *The $64 million* have to keep topping up energy for the wave shape to *answer* go anywhere. So if the wind doesn't keep blowing grains along, then the dune is nothing more than a stationary heap of sand. Likewise, if the drivers didn't step on the gas and use up fuel to drive forward again when they find themselves at the front of the traffic jam, then the traffic wave wouldn't progress backward up the road. Everyone would just sit there stationary on the road for days on end.

By comparison, a normal progressive wave traveling over the surface of the water happens to exist in a system that dissipates relatively little energy. Once the surface has been disturbed, that disturbance just travels along of its own accord as a wave.

At least, I think that's what he meant.

"Understanding things like traffic jams from a physical point of view is a totally new, emerging field of physics," Professor Sugiyama sympathized. "While the phenomenon of a jam is so familiar to us, it is still too difficult to truly understand why it happens."

Before saying goodbye, I was curious to learn how the professor had become so intimately involved in the minutiae of traffic behavior. Did his ideas come to him while he was stuck on the Japanese equivalent of the A303?

"Oh no," he told me, "I take the subway. I don't drive—I never have. I don't even have a driver's license."

～

All this congestion had left me wanting to clear my mind, so I decided to take another walk along the river. It was now late July, and towering Cumulonimbus clouds had brought a deluge of rain over the previous days. The waterway was up near the top of the banks and its normally languorous current had swelled to a brisk flow.

In the shade of a willow, a rock that normally protrudes from the surface was now submerged, but only just. As the current gushed over the obstacle, it made a pleasing babbling sound. This purl of passing water is the sound of cares being carried away.

And there, hovering on the surface in the lee of the rock, was our old friend the fluent standing wave. I'd never even have looked at it were my mind not brimming with thoughts of waves within currents. After all, there was nothing special about it. It was just the mundane rising and dipping of the current, as commonplace as the swish of the willow in the breeze.

But now that I *was* focusing, I began to feel that there was something, I don't know, *deep* about the way the standing wave hung there. There it was, a suspended moment in the current's progress. The water that rose and dipped to form the shape this second was lost downstream by the next. Yet the wave always remained in place, since more water was forever passing through.

It may sound like I was spending too much time on my own, but I began to wonder whether that wave hovering there in front of me was actually a "thing." It was certainly something that you can look at, point at, and think about. But it was also nothing more than a deviation in the current. It seemed such an abstract example of a wave. In fact, this whole business of waves within flows was difficult to pin down. Cluster solutions, dissipative systems, critical densities . . . it seemed that the more syllables the terms had, the more abstract and ungraspable they were. I was in a state of wave confusion. These waves had become so abstract that I needed to speak to someone who could help me back onto solid ground.

On returning home, I looked up my philosophy professor from university to ask if he knew of any philosophers who had something to say about the meaning of a wave in a stream.

Someone called Heraclitus was the guy, apparently, a Presocratic Greek philosopher who lived around 500 BC.

～

There's no evidence that Heraclitus wrote a single word on the subject of standing waves; nor on any other sort of waves, for that matter. He did, however, have a bit to say about rivers and currents. At least, there are a few sentences attributed to him on these subjects, even though none of Heraclitus's books have survived. All we know of his philosophy comes from others—a quotation here, a paraphrase there—just literary fragments, always interpreted by a secondary author.

The quotes that we do have reveal Heraclitus to have been someone who enjoyed being oblique, paradoxical even.* He reveled in self-contradictory aphorisms, such as "The road up and the road down are one and the same," "The beginning and the end are shared on the circumference of a circle," and "Living and dead are potentially the same thing, and so too waking and sleeping, and young and old; for the latter revert to the former, and the former in turn to the latter."[11] No wonder he was known by some as "Heraclitus the Obscure" and by others as "the Riddler."

For Heraclitus, *everything* was in flux. Although a flame appears to be a thing, something that is bright and flickers, it is in fact a

---

* This didn't serve him too well in the end. According to the biographer Diogenes Laertius, he died at the age of seventy in rather bizarre circumstances. Having developed dropsy, which resulted in the puffy build-up of fluids under the skin around his eyes, he asked the doctors in characteristically enigmatic fashion if they could "produce a drought after wet weather." Since they had no idea what he was talking about, Heraclitus decided to self-medicate by sitting in a cowshed and covering himself with dung. Apparently, he believed that the warmth would cause the fluid to evaporate out of him. There in a shed, covered in shit, is where he died.

process, a visible stage in something changing from one state to another. The same could be said of a river; Heraclitus claimed that "You cannot step twice into the same river; for fresh waters are ever flowing in upon you."[12] For him, an oak beam holding up the roof might seem to be something pretty permanent, *Help from* but this is just because its change is not obvious to us. *the Riddler* The beam will change over hundreds of years, just as a rock will disintegrate to sand over millennia. In summarizing Heraclitus's thinking, Aristotle wrote, "all things are in motion all the time, even though . . . this escapes our senses."[13] Heraclitus also pointed out that "the sun is new every day,"[14] which seems quite sensible now that we know it to be a colossal nuclear reaction emitting energy as it burns, and so in a state of continual change.

According to the rather more recent great philosopher Bertrand Russell, science has always "sought to escape from the doctrine of perpetual flux by finding some permanent substratum amid changing phenomena."[15] In other words, scientists peer down their microscopes in the hope of finding what remains fixed amid all the change that surrounds us. First, atoms were considered the indestructible building blocks—until, with the discovery of radioactivity, it was found that they could disintegrate. For a while, electrons and protons, which make up atoms, were deemed the things that never changed—merely rearranging themselves to form different substances. That worked until it was found that, when smashed together, these subatomic particles could disintegrate into pure energy—huge explosions of electromagnetic waves. This meant that, in the quest for permanence, the only thing left standing was energy.

Which must be why I found this little standing wave so intriguing. There, quivering in the current, it was unmistakably a wave, and yet it was nothing more than a continuum of passing moments, a stage in an ongoing process, a rising and falling diversion in the water's passage. If everything around us is just energy—the very embodiment of change, whose name, after all, derives from the Greek word for "active"—then that unassuming wave was a glistening symbol of the evanescent nature of all that we consider permanent.

The Fifth Wave

WHEN WAVES TURN NASTY

Sergeant David Emme was manning his vehicle's machine gun when the blast ripped through the side panels.

It happened on November 19, 2004, in the town of Tal Afar, northwestern Iraq, when the thirty-two-year-old American was part of a convoy transporting new recruits for the Iraqi Police Force. No sooner had the convoy set off than Sergeant Emme sensed something was wrong. Tal Afar was uncharacteristically quiet. The young children, normally running about and shouting in the dusty streets, were nowhere to be seen. Just a few teenage boys on a street corner; one of them looked at Emme and gave a cut-throat sign as the convoy passed.

Sergeant Emme used his radio to tell the rest of the convoy what he'd seen, warning that everyone should keep an eye open because he felt something bad was about to happen. As they negotiated the next roundabout, it did.

It was an improvised explosive device, or IED, hidden at the roadside on the left, that detonated just as they drew level. An *The meanest* expanding wall of ferociously high pressure raced ahead *wave of all* of an incendiary ball of expanding gases, carrying shrapnel within it that tore into the side of Emme's truck. Emme himself was exposed to the full force of the "shock wave," the most brutal of all types of wave.

When the sergeant came to within the vehicle, he was unable to see or hear properly. Shrapnel had entered his left eye and the blast had completely blown out his left eardrum. Some twenty-five insurgents, who had been hiding in nearby buildings, now opened fire on his truck and the rest of the convoy.

The next thing Emme knew, he was being pulled out of the truck by his driver and dragged toward a Stryker armored personnel carrier. Bullets snapped at their feet, and rocket-propelled grenades were exploding around them. Another sergeant had taken up Emme's position behind the damaged vehicle's machine gun and shot an insurgent barreling toward them in a car bomb, moments before he reached the stranded convoy.

Now within the relative safety of the armored vehicle, Emme and his driver were rushed to the Forward Operating Base at the edge of town. Having managed to walk down the ramp of the vehicle to waiting medics, here he lapsed into a coma. He was evacuated to a combat support hospital in Balad and then onto Baghdad. Besides blowing out his eardrum and injuring his eye, the blast had also fractured his skull and caused severe bruising of the left side of his brain. Neurosurgeons in Baghdad performed a craniectomy, removing a large piece of skull from the left temporal region to give Emme's brain room to swell. He woke, ten days later, in the intensive-care unit of the Walter Reed Army Medical Center in Washington, DC.

There, Emme would mistake nurses for CIA agents, believing he was back in Baghdad. His speech was nonsensical, and he was unable to follow simple commands. He suffered difficulties with his reasoning, memory, and problem-solving abilities, which only gradually began to return to normal after five months of cognitive therapy. His was a classic example of what has been dubbed the signature injury of the Iraq and Afghan conflicts: traumatic brain injury (TBI) caused by the shock waves of improvised roadside bombs.[1]

Besides the immediate cognitive problems suffered by returning veterans with injuries like Emme's, more general, longer-term symptoms are also common. Often described as post-traumatic stress disorder, these symptoms include anxiety, depression, and alcoholism, and are associated with increased rates of suicide. A 2008 study of veterans of Operation Iraqi Freedom in Iraq found that those showing symptoms of post-traumatic stress disorder three to four months after returning from combat were significantly more likely than the other veterans to have suffered the blast wave from a close-proximity bomb.[2]

A 2008 study by the RAND Corporation[3] found that of the 1.64 million U.S. troops deployed since 2001 in the Iraq conflict and in Operation Enduring Freedom in Afghanistan, a staggering "320,000 individuals experienced a probable TBI during deployment." That is almost one in five having suffered from a close-proximity blast. The reason the figure is so high is not just because the insurgents plant a lot of IEDs; ironically, it is also due to the effectiveness of modern Kevlar body armor. By protecting troops from shrapnel wounds, it ensures that far more survive close-proximity blasts than in previous conflicts. In the past, many more of these soldiers would simply have died.

The effect of a shock wave on the brain can be extreme even if there is no sign of physical damage. The blast creates a violent increase in air pressure, followed by a sudden decrease that produces a brief but ferocious "blast wind" that can reach speeds of 800mph, about ten times as fast as the winds of a hurricane.[4] These severe pressure shifts actually distort the skull, producing concussion or bruising of the brain tissue. Air bubbles can also form in blood vessels and travel to the brain, killing regions of the cerebral tissue. Computer simulations of these intense shock waves suggest that they are channeled down the air gap between the head and the helmet, subjecting the skull to a rippling distortion as they tear by. The effect on the soft brain tissue is akin to a head impact in a violent car crash.[5] The Kevlar helmet may protect the soldiers' heads from flying debris, but they offer little defense against the shock waves.

The effects of shock waves on troops are rarely fully appreciated. They often occur without any lasting visible wounds and many soldiers are reluctant to report mental problems following deployment for fear that doing so will harm their careers. The shock-wave injuries are not just a characteristic of modern warfare; they are in danger of becoming a hidden epidemic.

⌒

So what are shock waves?

Rather than being a *category* of wave, like ocean waves, electromagnetic waves or acoustic waves, shock waves are more like these waves when they are in terrible moods. In other words, any of these categories of wave can be described as shock waves when they are so intense that they behave in a very different way from normal, "mild-mannered" waves.

The most dramatic and noticeable shock waves are the acoustic ones. The pressure waves racing out from explosions are acoustic shock waves—particularly violent examples of the same compression and rarefaction waves that we sometimes hear as sound. But I'm just going to call them pressure shock waves or, when they pass through solids, density shock waves, in order to be less confusing, since most people find the "acoustic" term inextricably associated with normal sound waves, rather than the sort that would blow your eardrums out.

This good mood/bad mood analogy is hardly an accepted way of distinguishing shock waves from normal waves, but normal waves can develop into shock waves and vice versa, so moods seem a fair analogy to me. What, then, does it mean for waves to be in a furious mood? Regardless of the category they belong to—whether ocean waves, pressure waves, or whatever—shock waves will tend to exhibit one or more telltale characteristics:

They will have distorted expressions, compared to their mild-mannered brethren. The shape of shock waves is often different, as they lack the neat, symmetrical form of normal waves. A pressure shock wave, for instance, will typically arrive through

Invisible shock waves from the blasts of a battleship's 15-inch guns are revealed by the impression they leave on the ocean surface.

air toward you as one or more extremely steep jumps in pressure, followed by a more gradual return to normal air pressure. Mild-mannered pressure waves, by contrast, have symmetrical shapes of rising and falling air pressure.

They are usually in a tearing hurry. Shock waves generally travel faster through whatever medium they are in than the normal speed of their category of wave. While mild-mannered waves often have a fixed speed through a particular medium, regardless of their frequencies or intensities, shock waves travel faster than the others. And the more intense they are, the faster they rush.

They're too angry and impatient to obey the usual Ways of the Wave that the others do. Shock waves often don't reflect, refract, or

169

diffract in the same way as mild-mannered ones, nor when two shock waves overlap each other do they add and subtract in a straightforward way.

They tend to smash the place up. Shock waves often have a lasting, even damaging, effect on whatever medium they pass through. For instance, when a density shock wave passes through a solid it might break it to pieces. And if the medium is a liquid or a gas, it will tend to heat it up, sometimes severely. Mild-mannered waves, by comparison, tend to be polite enough to leave their surroundings as they were before they came.

Of course, physicists have a rather more robust way of defining shock waves than talking about mood swings. Anyone not of a scientific inclination might want to look away now, as I am about to summarize it. They say that a shock wave behaves in a "non-linear" way, while normal waves behave in a "linear" way, linearity being defined in terms of whether the wave obeys the "principle of superposition." This is the principle that the result of two waves overlapping each other is determined by a straightforward addition of their crests and troughs.

OK. You can look back now. It's all over.

～

Explosions are certainly the most obvious ways that shock waves are created. But they do not have to be man-made explosions. *Volcanic shock waves* For instance, an atmospheric shock wave emanated from the enormous volcanic eruption that blew the top off the Indonesian island of Krakatoa in 1883. It took ten hours and twenty minutes to travel the 7,220 miles to London. There it was recorded as a sudden jump in air pressure on meteorological barographs at the Greenwich Observatory, followed by a sudden dip and then a steady return to normal pressure.[6]

Keen wave watchers might have already worked out that traveling that distance in ten hours and twenty minutes suggests that the pressure waves traveled rather slower than they normally would. After all, the speed of sound (which is the speed at which pressure waves travel) through air at 40°F is about 750mph, while 7,220 miles in ten hours and twenty minutes amounts to about 700mph. This would seem to contradict the second characteristic of shock waves, on account of traveling slower, rather than faster, than a wave normally would. But in fact it would have bounced up and down within the atmosphere, making the distance of its journey much greater than how the crow flies.

Another natural source of shock waves is lightning. Whenever you hear the sound of thunder you are listening to shock waves. Here, the explosion is the violent expansion of the air caused by the incredible heat of the lightning bolt. This is what causes pressure waves with extremely abrupt fronts, followed by more *Thunder and* gradual returns to normal pressure at their backs. So is the *lightning* crash of thunder how we hear such a shape of pressure wave? This is what I asked when I spoke about thunder to Professor Mark Cramer, a shock-wave specialist at Virginia Tech. He told me that our ears register a jump in pressure like this as something like "the snap associated with an electrical spark, like when you attach cables to your car battery, or the clack of billiard balls."

So why, I asked, does thunder sound more like an enormous crash and deafening, tearing sound (at least, when it is close by)? This, he pointed out, is partly because "what we see as a single lightning bolt is really multiple discharges." With multiple shock waves following each other in quick succession, the sound is extended. The same result comes from the fact that "thunder is generated all along the lightning bolt, which is typically several miles long." This means that shock waves originate at different points along its branching path, with the more distant ones taking longer to reach you than the closer ones. Both these factors mean that you do not hear the electric-spark click of a single shock wave, but the ripping sound of multiple ones joined together.

Besides having the abrupt fronts of shock waves, the thunderous pressure waves caused by flashes of lightning also travel faster than normal sound waves. In fact, they follow the shock-wave rule that the more intense ones—the pressure waves that sound louder—travel faster than the less intense, quieter ones. This is one reason, explained Cramer, why the thunder from a storm many miles away is a deep rumble, rather than the high, tearing sound of close by. "The wave speed is dependent on amplitude," he explained, the amplitude of a pressure wave being another way of describing its intensity, or volume. "Thus, different bits of the wave will propagate at different speeds, leading to a distortion of the wave." This distortion, I learned, helps explain the change in sound of thunder with distance.

An explosion, whether caused by a bomb or by the rapid expansion of air along a lightning bolt, produces a confusion of frequencies and intensities—shrill sounds and booming sounds, louder sounds and quieter sounds—all mixed in together. Near the lightning bolt, these pressure shock waves combine into a deafening, tearing sound. But farther away, the sound becomes drawn out. Since the louder, more intense shock waves travel faster than the quieter, less intense ones, they race ahead. Thus, the succession of shock waves becomes more spread out the farther they travel, making them sound deeper in combination. This is a little like the pitch of sound you hear when you run a stick along an iron railing: if you run it fast, all the individual clangs join to make a higher-pitched noise than if you run it slower. The clangs are the same, but we hear them join together into a higher or lower noise, depending on how fast they follow each other. This drawing out of the succession of shock waves is one reason why there is such a pronounced difference in tone of thunder with distance.*

Fortunately, you don't need to experience a bomb blast or a lightning strike to feel a shock wave at first hand. In fact, you cause one whenever you tuck your knees to your chest and dive-bomb into a

* Another reason is that more of the sound from the bolt will have bounced off buildings and hills before reaching you, causing an echoing reverberation. And then there is the fact that high-frequency sounds decay more rapidly than low-frequency ones, so more of the high sounds will have been dampened by the time the shock waves from a distant storm reach you.

Feel free to add the sound of the shock waves yourself.

swimming pool. The turbulent wall of water thrown out as you hit the surface is the aquatic equivalent of the crash of thunder near a lightning bolt.

Competitive swimmers are not big fans of shock waves generally. They try to avoid them as much as possible, since they're a waste of energy. This is one reason why you don't see them doing whopping great dive-bombs when the gun goes off for the 200m front crawl. (Although I rather wish they did.) Instead, they aim to produce the smallest possible splash as they dive into the water, pointed fingers first. This way, they minimize the water resistance and thus the shock wave produced by pushing the water out of the way. A reduced shock wave means less energy lost to the water. Once they are off, however, another equally unwelcome aquatic shock wave rears up. This is the "bow wave," just in front of the swimmer's head, as he or she speeds through the water.

Wait a minute. A bow wave? Where's the crash and boom—or even the splash—that we've come to expect of a shock wave? Actually, although explosions are the most obvious ways shock waves are created, they are not the only ones. They can also result *A puny sort of* from the bow waves that spread from the front of a moving *shock wave* object—be it a swimmer, or a boat on the water, or some object traveling through the air—so long as it is traveling fast enough through the medium. A bow wave only develops into a shock wave when the object is going at least as fast as waves naturally travel through the medium so that they can't get out of the way fast enough and end up bunching together on top of each other.

If the bow wave of a swimmer sounds rather mild for a shock wave, how about another shock wave produced in this way: the sonic boom created as a jet flies at supersonic speeds.

This is another example of a shock wave of air pressure—one that we can hear as sound—rather than a shock wave on the surface of the water. But the principle is just the same. As it flies through *That's a bit* the sky, a plane will produce pressure waves at its front and *more like it* back, since it displaces the air, which has to move aside rapidly at the front and rush back in behind, as it once more fills the space left in the plane's wake. These sudden movements of air cause jumps and dips in the air pressure, which spread out from the plane as pressure waves. These pressure waves expand as spheres, some emanating from the nose, some from the tail. They are formed whatever speed the jet is flying at. Although we often hear pressure waves as sound, we don't normally hear these because the roar of the jet engines drowns them out. Unless, that is, the jet is traveling at, or faster than, the speed of sound. Then we most definitely do hear them.

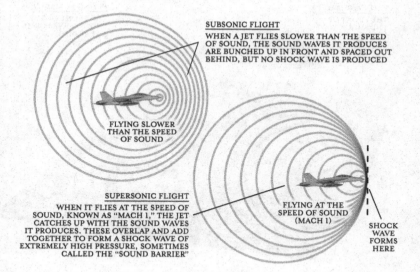

SUBSONIC FLIGHT
WHEN A JET FLIES SLOWER THAN THE SPEED OF SOUND, THE SOUND WAVES IT PRODUCES ARE BUNCHED UP IN FRONT AND SPACED OUT BEHIND, BUT NO SHOCK WAVE IS PRODUCED

FLYING SLOWER THAN THE SPEED OF SOUND

SUPERSONIC FLIGHT
WHEN IT FLIES AT THE SPEED OF SOUND, KNOWN AS "MACH 1," THE JET CATCHES UP WITH THE SOUND WAVES IT PRODUCES. THESE OVERLAP AND ADD TOGETHER TO FORM A SHOCK WAVE OF EXTREMELY HIGH PRESSURE, SOMETIMES CALLED THE "SOUND BARRIER"

FLYING AT THE SPEED OF SOUND (MACH 1)

SHOCK WAVE FORMS HERE

How a shock wave is formed by flying at the speed of sound.

When a jet reaches the speed of sound, known as Mach 1,* it is traveling at the same speed as the pressure waves it produces. This means that the waves created at the jet's nose cone are unable to move ahead of it since the plane keeps pace with them. The pressure waves therefore pile up, each peak of pressure combining with the previous ones, each adding onto the others to form a pressure bow wave of greater and greater intensity. When the jet flies at the speed of sound, the pressure waves combine to form a shock bow wave. An abrupt shock wave of increased air pressure extends outward from the front of the plane, as a bow wave moving with it at the speed of sound, while a shock wave of decreased air pressure, more of a "stern wave," really, extends from the tail.

On the ground, exploding sounds can be heard as the high- and low-pressure shock waves pass over with the screaming jet. The one

....

* The reason we use the terms "Mach 1" and "Mach 2" rather than specific speeds is because the actual speed of sound changes depending on the air temperature. At 32°F, Mach 1 is about 740mph, while at −4°F, it is more like 714mph.

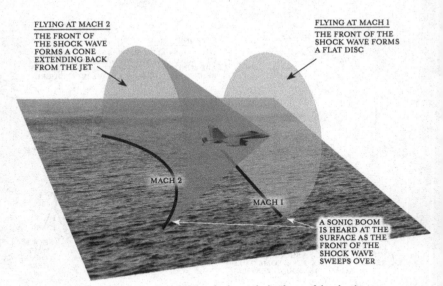

FLYING AT MACH 2
THE FRONT OF
THE SHOCK WAVE
FORMS A CONE
EXTENDING BACK
FROM THE JET

FLYING AT MACH 1
THE FRONT OF THE
SHOCK WAVE FORMS
A FLAT DISC

MACH 2

MACH 1

A SONIC BOOM
IS HEARD AT THE
SURFACE AS THE
FRONT OF THE
SHOCK WAVE
SWEEPS OVER

As the jet travels faster than the speed of sound, the front of the shock wave
changes from a flat disc to a cone.

at the front is immediately followed by the one at the back (though
the two bangs are usually too close to distinguish unless the jet is
flying high). But the experience is very different for the pilot, who
won't hear a boom when traveling at Mach 1 since the front of the
shock wave always remains just ahead of the nose of the jet. The
pilot only hears the boom of the shock wave when he increases
the thrust enough to fly faster than the speed of sound, thus
breaking through the "sound barrier," which is simply the shock
wave of high pressure just ahead of the cone. It is described as a
barrier because a considerable amount of extra thrust is required to
pass through this region of increased pressure. If the speed of sound
in the air through which the jet is traveling is 740mph, the increase
in thrust required to accelerate from 740 to 750mph will be a lot
more than it was from 730 to 740mph, since this involves traveling
faster than the speed of sound, faster than Mach 1, and thereby
overtaking the region of high pressure that is the shock wave front.
As the pilot breaks through the sound barrier, the region of high
pressure passes over the cockpit, sounding like a boom.

With increasing speed over Mach 1, the position of the shock-wave front changes. At Mach 1, it extends outward at the front of the jet, like an enormous plate of high pressure stuck onto the end of the nose cone, and another of low pressure stuck onto the tail. Having broken through the sound barrier, both plates change into cones extending backward from the nose and tail. At Mach 2, twice the speed of sound, these shock-wave cones are at an angle of 45°. Were a supersonic jet to pass overhead at such a speed, you would not hear any sonic booms until after the aircraft had passed you and the shock-wave cones extending behind it arrived.

When air, or any other gas, is compressed it heats up; when it expands it cools. And this is why the shock wave produced by a supersonic jet can sometimes be rendered visible as a ghostly, transient cloud, known as a "shock collar" or "shock egg." The very high pressure at the front of the shock wave has a region of low pressure just behind it. This drop in pressure can cause the air to cool so much that the water vapor it contains momentarily condenses into droplets of cloud. Depending on its speed, this supersonic cloud appears as a disk (Mach 1) or a cone (faster than Mach 1) attached to the fuselage of the jet.

It can look a bit like one of those mints they bring you in restaurants with your bill (see next page).

The sounds of the supersonic shock waves vary too, depending on how high the jet is flying. From very high altitudes, each shock-wave cone (front and back) is spread out by the time it reaches the ground, and its booms sound deep, like loud thuds. But a low fly-by produces a sharper sound, like a rapid pair of gunshots or even the cracks from a lion tamer's bullwhip. (Actually, it would have to be the sound of two, coordinated lion tamers, who could crack their whips almost in unison.) Whether thuds or cracks, these shock waves are, of course, quickly followed by the ferocious roar of the jet's engines.

～∕e

The comparison between supersonic jets and bullwhips is not a fanciful one. After all, the sound of the whip also starts life as a

An F/A-18 Hornet (TOP) and Super Hornet (BOTTOM) break the sound barrier, while flying through very large mints.

shock wave, caused by supersonic movement. The terms a "cattle-man's crack" and a "coachman's crack" are not the cowboy and carriage-driver equivalents of a "plumber's butt." They are, in fact, techniques designed to make the tip, or "cracker," at the very end of a bullwhip move faster than the speed of sound, thereby producing the snap of a shock wave. The experienced cattleman or coachman can crack his whip with what seems a very relaxed sweep of the hand. He sends the energy down the length of the braid in the form of a traveling loop, and pulls back on the handle to increase the tension as it goes. Initially rather leisurely, this loop is converted into a violent, supersonic crack at the far end of the whip by the way it becomes increasingly narrow and flexible down its length.

Rawhide shock waves

The "thong," the section nearest the handle, is the thickest part, made of many leather cords braided together. This gradually tapers, becoming thinner and thinner until the point where it is connected to the "fall": a single, flexible leather cord. At the end of this is tied a small piece of very flexible string, nylon, or wire, which is the cracker.

Just as a narrowing, shallowing river channel can focus the energy of an incoming tide into a steeper, faster wave front, so the whip's taper concentrates the energy sent down it as a traveling loop or wave shape into an even smaller volume. Mathematicians at the University of Arizona worked out that if the diameter of the cracker at the far end of a typical 6ft whip is one-tenth that of the thong at the handle end, then the whip will concentrate the energy of the whip's movement so that a loop at the tip travels thirty-two times as fast as its initial speed near the handle.[7]

So it is easy, with practice, to make the loop travel at supersonic speed. High-speed photography has shown that for the loop to travel at the speed of sound as it reaches the end of the whip, thereby producing the shock-wave sound, the cracker itself ends up traveling at *twice* the speed of sound. Yeee-ha!

Does my cattleman's crack look big in this?

Remember that, besides their sudden arrival, on account of their abrupt fronts, and the fact that they tend to travel faster than waves normally do, the third characteristic of shock waves is the considerable, sometimes damaging, effect they have on whatever they pass through. Sergeant Emme found that a shock wave can leave a lasting impression, as it did with the severe bruising it caused as it passed through his skull and brain. Of course, some soldiers unlucky enough to be in close proximity to explosions die from the shock waves passing through their bodies.

As they pass through air, the extremely high pressure at the front of some shock waves can greatly increase the temperature of the air they pass through. In extreme cases, the air can be heated so much that it changes chemically. This is precisely the effect a shock wave had on the air at the outer reaches of the earth's atmosphere, resulting in one of the most heart-stopping moments of drama in the history of space flight. This occurred during NASA's ill-fated *Apollo 13* manned lunar mission. You'll have to hang in there to see the part played by the shock wave, as it didn't take center stage until the closing minutes of the flight.

The world's media had shown little interest following the launch on April 11, 1970. After all the fuss made about the moon landings the previous year, *Apollo 13* promised to be just more of the same. Once things started to unravel, however, the *Just another* mission became front-page news across the globe. Every *lunar mission* stage of the unfolding drama was played out on live television—it was as if the whole world was holding its breath to see whether the three astronauts on board would make it back alive.

The problems began two days into the flight, when a member of the crew was conducting the routine task of turning on the propellers that stirred the liquid oxygen within the craft's two supply tanks. As he did so, a spark from an exposed wire within one tank caused it to explode, ripping a hole in the side of the service module (the unmanned section of the spacecraft housing its propulsion, electrical, and air-conditioning equipment) as well as damaging the other oxygen tank.

Hearing the bang, the astronauts knew that something had gone seriously wrong. With the warning lights flashing, power systems failing and their instruments going haywire, however, they weren't sure quite what. Then Commander Jim Lovell glanced out of the window and saw that their craft was hemorrhaging oxygen into space. This was when he radioed Mission Control with the famous words, "Houston, we've had a problem."

Aborting the moon landing, the experts back at the Johnson Space Center focused on the task of returning the crew safely. The three astronauts were instructed to move from *Odyssey*, the stricken command module that would normally accommodate them during the earth take-off and landing and while in orbit, and over to *Aquarius*, the lunar module that was designed for the moon landing and take-off. Since there would now be no lunar descent this would have to serve as a space-age life raft. But the limited oxygen of the lunar module would last, at most, forty-five hours, nowhere near enough to sustain them until the scheduled landing. So the scientists in Houston had to calculate a risky new trajectory to accelerate their homecoming. They decided to use most of the lunar module's fuel to shift the return trajectory around the back of the moon and reduce the flight time by nine hours. If everything worked out right, the moon's gravity would send the spacecraft, like a slingshot, shooting back to Earth. If it didn't . . . well, there was no Plan B.

Once they had managed to set themselves on the right course, the astronauts had to disable the computer navigation, guidance and heating equipment to conserve their dwindling power. All they had left was the radio with which to talk to Earth, and the fan system that circulated the air. Without heating, the temperature eventually dropped to around 39°F.

It was the filter units designed to remove from the air the toxic CO_2 exhaled by the astronauts that were to become their next big problem. The ones accessible in the lunar module were designed to work for only a few hours—plenty of time for a lunar landing, but not for the journey back to Earth. With CO_2 levels becoming critical, the eggheads had to devise a way of adapting the spare filter units from the command module using the duct tape and bits of plastic and cardboard that the astronauts had on hand. Then, as if this weren't enough, Mission Control told them that they were heading back to Earth at too shallow an angle. They were in danger of missing the planet altogether and being flung off into an enormous orbit from which they could never return. The astronauts were instructed to use the lunar module's descent propulsion system to correct their trajectory manually, lining up Earth through the module window to ensure they were heading in the right direction.

Even when back on course, the tension did not let up. They would have to return to the command module and jettison the lunar *OK, I'm getting* module life raft for their descent to Earth. But no one knew *to the point* whether the explosion had damaged the heat shield, which was designed to protect them as they re-entered the atmosphere. And here, finally, is where the shock wave came in.

What they *did* know was that, as the command module hurtled toward Earth at 25,000mph, an intense bow wave would form in the air at its front. The pressure would become so great that it would heat the air to 4,900°F. They also knew that such intense heat would change the state of the air from gas to "plasma," which is when it becomes so hot that electrons are ripped away from the air atoms.

The fiery bow wave at the front of the command module offered a vivid display of the violence of shock waves, and how their force can modify the material they pass through.

The astronauts and Mission Control were aware that the loose electrons in the plasma would make it a very good conductor of electricity and so would block the electromagnetic waves used for communication with the capsule on re-entry. In practical terms, this meant that there would be a radio blackout for the three minutes or so that the speed was high enough for the shock wave to create the plasma.

The big unknown was the state of the heat shield, upon which the astronauts' lives depended. Was it damaged? Would it hold up in the extreme temperatures of the shock wave?

The command module entered the outer atmosphere, the pressure of the shock wave heating the air into a radio-wave-blocking plasma. Communications between *Odyssey* and Mission Control were severed and news anchors around the world explained to their rapt audiences that there was now nothing to do but wait.

After three minutes, Mission Control tried to contact the craft: "*Odyssey*, Houston. Standing by, over." No reply.

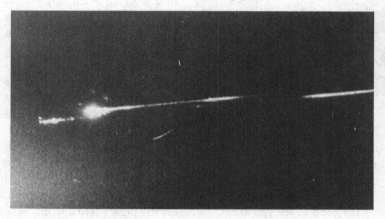

A passenger aboard a flight from the Fiji Islands to New Zealand happened to snap this photograph of *Apollo 13*'s jettisoned service and lunar modules burning up on re-entering the earth's atmosphere. Might the same fate befall the damaged command module with the three astronauts on board?

BEFORE

DURING

AFTER

The *Apollo 13* lunar mission, as experienced at Mission Control. Dramatic tension courtesy of a re-entry shock wave.

Three minutes became four, and still there was no radio signal from the astronauts.

Rescue helicopters hovered in place at the expected splash-down point southeast of American Samoa in the middle of the Pacific Ocean. The eyes of all the controllers in the command room of the Johnson Space Center were glued to their stopwatches. After four and a half minutes, some were fearing the worst.

Then, with a crackle of static, the voice of the pilot, Jack Swigert, came over the radio. It must have felt as if the whole world breathed a sigh of relief.

In a reversal of the usual time shifting when suspenseful moments are dramatized for the big screen, Ron Howard's film *Apollo 13* actually *shortened* the duration of the shock-wave-induced radio blackout. There was no need to draw it out to raise the tension. Reality had out-Hollywooded Hollywood.

Artistic license crept in, however, when it came to the script at this critical moment. "Hello, Houston. This is *Odyssey*," said Swigert in the movie, as the orchestral score swelled to a crescendo. "It's good to see you again." Whereas the pilot's actual words were a decidedly un-Hollywood, "OK, Joe."

I may have given shock waves a bit of bad press. The destructive effect they can have on a medium they pass through can, in fact, be rather good news. Particularly if you are suffering from a kidney stone.

"Extracorporeal shock-wave lithotripsy" is the medical term for the non-invasive use of shock waves to break down the hard crystalline deposits that can develop in kidneys. The patient lies on a special couch fitted with a shock-wave generator that *Pulverizing* focuses high-intensity sound waves onto the kidney stone. *kidney stones* Key to the procedure is the fact that the energy of the shock waves is absorbed most by the body at abrupt changes in density. As they pass from the soft kidney into the hard crystal and out the other side, they cause stresses in the stone that break it down. An hour's treatment, during which up to eight thousand shock waves are administered, is enough to pulverize a typical stone, say ¼ or ½in across, into particles small enough to be peed out. Nice.

To avoid collateral damage to bones and cartilage, the shock waves must be focused at the right spot. This is where waves of a less aggressive nature play a supporting role. The operator locates the exact position of the kidney stone, using either an ultrasound or a real-time X-ray scanner, neither of which uses shock waves to form an image. The ultrasound scanner sends out benign, high-frequency acoustic waves, forming an image, much like a submarine's SONAR, from the echoes that bounce back. (You could say that the kidney stone is the underwater mine about to be torpedoed by the pressure shock waves.) The X-ray scanner, known as a "fluoroscope," sends low-intensity, high-frequency electromagnetic waves through the patient. Solid objects like kidney stones scatter and absorb the X-rays more than soft tissue, so they appear as shadows on the detector at the far side.

◠

At a rather larger scale, the earth also has a solid lump within the softer "tissue" of its own body. Dig down some 3,200 miles, which takes you about three-quarters of the way to Earth's center, and there you would find it.

We know a surprising amount about the inside of the earth. The outer layer is a hard crust, with an average thickness of around 20 miles, beneath which lies a solid mantle of differing rock, which *The innards of* is rigid to around 40 miles, and highly viscous below. At *the earth* a depth of around 1,800 miles begins the outer core, a liquid layer of molten iron and nickel, the currents within which are believed to cause the earth's magnetism. And there, right in the middle of this liquid, is the solid inner core, which is about 1,500 miles across. It is almost three-quarters the diameter of the moon, and is made of solid iron and nickel (rather than cheese).

How do we know all this, given that the deepest man has ever dug is 7.6 miles? (This is what the Russians managed below the Kola Peninsula in the Arctic.) The answer is through the shock waves caused by earthquakes.

The World-Wide Standard Seismograph Network, completed in 1961, was a global system of recording devices that were designed to listen for the rumbling shock waves of nuclear explosions. This international monitoring system was intended to ensure compliance with the 1963 nuclear test ban treaty agreed between the United States, UK, and Soviet Union, which forbade the above-ground testing of nuclear weapons.

This worldwide web of seismographs also began recording earthquakes to new levels of accuracy. Triangulation could be used to pinpoint each quake by comparing the arrival times of the first seismic waves at different instrument locations. For the first time, it became clear that there was an order to the distribution of the earthquake epicenters—the points on the surface, directly above the underground focuses, where the ruptures of the earthquakes start. Far from being positioned randomly, these were concentrated along well-defined fault lines. This discovery led to a revolution in our understanding of the earth's crust, confirming the theory of "plate tectonics," which states that the surface is made up of enormous, rigid plates, gradually moving relative to one another. Earthquakes occur most at the boundaries, due to the sudden release of friction on a monumental scale. The cataclysmic shearing of the tectonic plates produces shock waves that reverberate through and around the planet. Indeed, of the daily fifty or so earthquakes worldwide that are strong enough to be felt locally, a handful produce vibrations that are so powerful they can be detected by modern seismographs anywhere on the globe.[8]

Within the ground an earthquake generates a mess of different waves, all of which behave in slightly different ways. Seismic waves fall into two main categories, according to whether they travel through the body of the earth, or just around its surface.

Waves that travel through the body of the earth move more rapidly than the surface ones. They are the waves that seismologists use to determine the position of earthquakes, by comparing their arrival times at different monitoring stations. The fastest, and so the first to arrive, are known as "primary waves," or "P waves," which zoom through the earth at between 5 and 8 miles per second.[9] They are longitudinal waves, consisting of compression and rarefaction of the earth's interior. In other words, they are density waves, push-and-pull waves, which travel by means of the earth vibrating to and fro along the direction of the wave's travel. P waves are shock-wave versions of those other compression-rarefaction waves: acoustic waves. And just as we can hear sounds under water as well

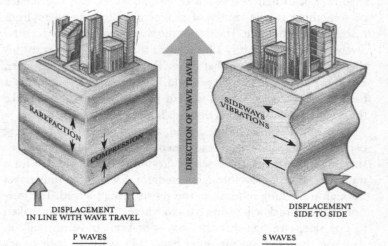

Seismic body waves travel through the planet's interior.
P waves are longitudinal, while S waves are transverse.

as through the wall from next door, so P waves can pass through liquid as well as solid parts of the earth's interior. The difference from other acoustic waves is their abrupt, violent fronts and the intensity of their vibrations.

Having first recorded the arrival of P waves, a seismograph located several thousand miles away from the focus of a large quake would then detect a different sort of wave, arriving a few minutes later. This is the other type of body wave, known as "secondary waves," or "S waves." They travel through the interior at about 60 percent of the speed of the P waves, from about 2.5 to 7.5 miles per second.[10] S waves are transverse waves, meaning that the rock shakes from side to side or up and down compared to the direction the waves are traveling.

By carefully examining the wave evidence immediately following an earthquake—which types of seismic waves arrive at which monitoring stations and at what times—we have been able to piece together a precise picture of the interior of the planet.

For instance, on the opposite side of the planet from a quake there is always an absence of S waves. This is like an enormous shadow, not of light waves but of seismic waves, and is evidence of something in the center of our planet blocking them from reaching the other side. The size of the shadow suggests this central S-wave-blocking region is a little larger than Mars. Thanks to this shadow, geologists know that below the earth's mantle is a layer, called the outer core, which is liquid. The reason they know this is because transverse waves—those that jiggle from side to side, like the S waves—can never pass through liquids.

Why not? Because, unlike a solid, a liquid won't tend to spring back from any side-to-side movements. It lacks the resistance to shearing movements, the essential "restoring force," needed for a transverse wave to pass through a medium. S waves can only travel through something solid, since this resists being sheared. The interior of a solid rock will spring back to where it started when shaken to one side, ensuring that the wave vibrations travel through it. Any sort of liquid, on the other hand, will refuse to spring back from side-to-side shaking vibrations, merely absorbing them by flowing back and forth. So a liquid layer within the earth's interior must be blocking the transverse S waves and causing the shadow on the far side of the planet.

By deductions like this, geologists have been able to build up a working model of just where the solid crust changes to the viscous mantle, then to the liquid outer core, and finally to the solid inner core. Further clues come from the way both P and S waves change speed as they pass through the differing densities of the earth's interior. This causes them to change direction, which is to say that they are refracted. They trace arcing paths as they speed through the gradual density gradients within layers, and make abrupt changes in direction at the boundaries. But piecing together the clues is no simple matter. As if to throw us off the scent, one type of wave can change into another: an S wave can change to a P wave and vice versa.

The network of seismographs acts like an enormous medical scan of the planet. Just as the absorption and scattering of ultrasonic waves sent through a woman's womb reveals the image of her unborn child, so the waves sent out by the shock of an earthquake enable us to peer into the belly of Mother Earth.

With all this talk of seismic waves, it might seem a major omission not to have mentioned their most significant feature, as far as we are concerned: that they can cause terrible destruction on the surface. But it is not actually the body waves that cause *The most* most of the damage. This is generally done by the surface *destructive* waves. As their name suggests, they are restricted to the *seismic waves* outer layer of the planet. They spread from the epicenter, traveling around the solid crust without passing through the interior. With slightly slower speeds than the S waves, they are usually the third type of wave to be picked up on seismographs.

There are two types of surface wave, "Love waves" and "Rayleigh waves," named after the scientists who mathematically described them. Love waves shake the surface horizontally: from side to side, compared to the direction of the wave. Rayleigh waves, on the other hand, roll up and down, with the ground moving in oval orbits. In this sense they are more like ocean waves.*

....

* Although the ground rotates in the opposite direction to the water in ocean waves.

The similarity between the rolling Rayleigh waves and those on the sea was noted by a witness to the earthquake that shook Charleston, South Carolina, in 1886:

> *The ground began to undulate like a sea . . . I could see the earth waves as they passed as distinctly as I have a thousand times seen the waves roll along Sullivan's Island beach . . . The waves seemed to come from both the southwest and northwest and crossed the street diagonally, intersecting each other, lifting me up and letting me down as if I were standing on a chop sea.*[11]

The degree of surface waves, and hence the quake's potential for destruction, depends not just on its magnitude, but also on the depth of its focus. Even a large earthquake, whose rupture begins over 200 miles down, will produce mild surface seismic waves, compared to one near the surface. The devastating, magnitude-7.0 quake that brought utter devastation to the Port-au-Prince region of Haiti in January 2010 was just 8 miles below the surface. Its

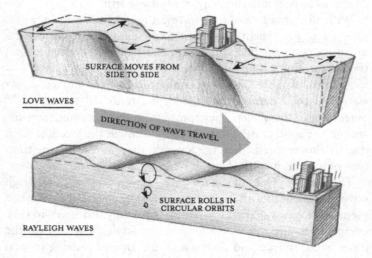

Seismic surface waves follow the surface of the planet and are more destructive than body waves. They are called Love waves when the ground shakes from side to side and Rayleigh waves when it rolls.

surface vibrations were therefore enormous, and almost completely flattened the city and neighboring towns. The buildings and infrastructure of this impoverished nation were entirely inadequate to stand up to them and those from the many aftershocks that followed. This particular series of waves killed an estimated 230,000 people in a matter of hours.

~℮~

One animal that is no stranger to shock waves is the little fellow shown on the next page.

The snapping shrimp, also known as the pistol shrimp, is a member of the *Alpheidae* family of crustaceans, found in the reefs of tropical and temperate oceans across the world. Typically no more than a couple of inches long, it might seem too puny to be messing with shock waves. But remember, size is not a defining characteristic of shock waves. They stand out from other waves on account of their abrupt fronts, their increased speeds and *Size isn't* *everything* their effects on whatever they pass through. Mini-shock waves not only exist, but are the tools of the trade for this pugilistic little shrimp. I say pugilistic because, with one claw much larger than the other, it reminds me of a boxer who has lost a glove.

By closing its monster claw at incredible speed, this shrimp produces a snapping sound that it uses to communicate. Dive near a colony of these shrimps and you will be reminded of an underwater popcorn factory. The collective sound is so noisy that it was found to disrupt the ability of submarines to listen for enemy subs during World War II.

The snapping sound is used not only for shrimp-to-shrimp Morse code, however: it has a more deadly purpose. When it snaps its claw, the shrimp produces a 65mph jet of water. The effect is a little like that hand-squirting thing that kids do in the pool by rapidly clenching their fists, albeit taken to an extreme. *Alpheidae* can use these jets of water to stun, or even kill, small fish and other species of shrimp. All very impressive, but, most surprising of all, it isn't the claw banging shut like a castanet that produces the snap or the stun effect, but an underwater shock wave.[12]

Back off. This claw's loaded.

Although the shock wave occurs on a tiny scale, it is a scaled-down shock to rival the eruption of Krakatoa. So fast is this jet of water that it produces a "cavitation bubble." In the jet's wake, the pressure drops so low that the seawater momentarily turns into a bubble of water vapor. Within milliseconds, that bubble implodes with such violence that it produces a shock wave that travels through the water as a violent increase in pressure that can stun prey at 1.5in.

The water vapor in the collapsing bubble heats up as it is instantly compressed back into liquid, reaching a temperature of around 8,500°F. Believe it or not, this is close to that ubiquitous measure of hotness, the surface of the sun, and causes a sudden flash of light. Lasting no more than a millisecond, it is too brief to be seen by the human eye but can be recorded on 40,000-frames-per-minute video. Although the phenomenon of a pressure shock wave producing light in this way is known as "sonoluminescence," the researchers who photographed it in the natural world like this dubbed it "shrimpoluminescence."[13]

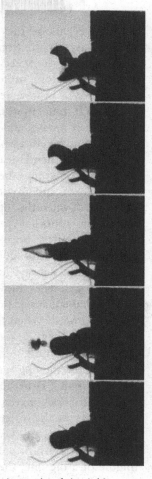

A snapping shrimp's big claw doing what its little claw can't: produce a cavitation bubble that forms a shock wave with which it can stun prey.

～

One of the striking things about waves is how little we notice them.

Oxymoronic as this might sound, there is some sense to it. We pay attention to the information that waves carry, but generally we have no need to notice their waviness—no need to be aware of the subtle, sprightly messengers. You don't have to understand how light works in order to see what is around you. And when someone tells you that they love you, the last thing on your mind, I would hope, is that their declaration takes the form of a succession of periodic pressure waves.

Which is another reason why shock waves stand out from the others. They show no such subtlety. For these relentless brutes of the wave world, the medium is the message.

As we've seen, shock waves can form in two ways: through explosive events—be they bombs, volcanoes, or shrimp claws—and as bow waves at the front of objects moving through a medium at, or beyond, the speed at which waves normally travel through it. But there is one more way that shock waves form: when one particular type of wave reaches the end of its natural life.

It is, in fact, the most familiar shock wave of all. Close your eyes and think of a wave, and you are probably picturing it.

Have you any idea what I'm going on about yet?

It is our old friend the ocean breaker. You'll remember that when a sea wave travels into the shallows near the shore, it bunches up as it slows down, and grows steeper and steeper until it becomes top heavy. Well, when the top of the wave comes tumbling over, it has turned into a shock wave.

Think about it: breakers show all the characteristics of shock waves. They have abrupt fronts—the tumbling, spilling, and surging front of water is always steeper than the more gradual slope at their *You've been* backs. They move faster than normal waves—at least the *watching them* dramatic ones do, as the water crashes down and shoots *all along* forward. When they are at the stage of tumbling, chaotic white water, they are too unruly to obey those refined Ways of the Wave: reflection, refraction, and diffraction. And much of the energy of the moving water is dissipated into the surroundings as heat and sound, rather than continuing as an ocean wave.

Like all shock waves, ocean breakers can be dangerous and destructive. As any surfer knows, to be in the wrong part of a large breaking wave at the wrong moment can be fatal, but for those who aren't foolish or unfortunate enough to be caught up in the middle of a big one, these are simply shock waves at their most beautiful. I could watch them for hours. The moment a normal wave changes to a shock wave is magical.

One minute it's an ordered undulation of the surface, the next it's a chaotic turbulence of air in water, and of water in air. Where else do you see that descent from order to chaos played out so gracefully, again and again, before your eyes? To call it a transition from linear to non-linear wave may be technically correct, but it doesn't exactly do it justice.

A wave breaking is a wave dying—or the moment when its aqueous life reaches an end, even if its energy lives on in other forms. And it is as a shock wave that it finally offers up its life force to the air and shore.

"Break, break, break," wrote Alfred, Lord Tennyson,

At the foot of thy crags, O Sea!
But the tender grace of a day that is dead
Will never come back to me.[14]

The Sixth Wave

There is a scene in the 1989 rom-com *When Harry Met Sally* when Harry Burns, the character played by Billy Crystal, is discussing his collapsing marriage with his friend Jess. They're at a football game, though they are paying no attention to it. The scene starts as the crowd around them are sitting back down from a stadium wave that has just gone by.

Harry's explaining how his wife has said that she wants a trial separation. He wears the blank expression of someone whose life is coming apart at the seams.

"So I say to her, 'Don't you love me anymore?' You know what she says?" Jess shakes his head. "'I don't know if I ever loved you.'"

"Oooh that's harsh," says Jess, just as a cheer rises from the crowd and another wave sweeps by. They stand with their hands in the air and participate in the wave, almost unthinkingly, before carrying on with the conversation.

Harry says that his wife had only just told him that she was going to move out to stay in a friend's apartment, when the bell rang and he opened the door to the moving men. "So I said, 'Helen, when did you make this arrangement?' She says, 'A week ago.' I said, 'You've known for a week and you didn't tell me?' And she says, 'I didn't want to ruin your birthday.'" Just then, another roar rises from the crowd, and everyone jumps up again. Automatically, with blank expressions, the two stand to join the wave, before continuing.

It turns out that Harry's wife has been lying to him, and that she's actually run off with a tax attorney. "Marriages don't break up on account of infidelity," Jess lectures. "It's just a symptom that something else is wrong."

"Oh, really?" says Harry, as a crescendo of cheers builds for one last sweep of the wave. "Well, that symptom is fucking my wife."

When I first watched the movie, I remember finding the scene funny because of the passive way the two characters were swept up in the wave. It seemed to echo Harry's inability to halt the tide of his foundering relationship. Of course, people aren't swept up in a crowd wave like this; they have to actively participate for it to travel around a stadium. Nevertheless, the wave does seem to have a life of its own. It feels more than the sum of its parts, rolling along on the collective energy of the crowd. It seems almost plausible that one might lift you up like a marionette as it swept by were you too shell-shocked to participate.

~c~

Known in the UK as a "Mexican wave," and in Latin America by the Spanish translation "*la Ola*," the stadium wave caught the attention of the world's media at the 1986 soccer World Cup, held in Mexico.

Capitalizing on these spontaneous crowd celebrations, Coca-Cola, ever the marketeers, were quick to associate themselves, running TV ads that showed the waves, and ended with the line "*Coca-Cola, la Ola del Mundial*" (the World Cup Wave). Since the company was one of the World Cup sponsors, these ads, and so the waves, were watched by a cumulative audience of 13.5 billion TV viewers.[1]

Most of us have contributed to a stadium wave at one time or another, whether at a large-scale sporting event or a Coldplay gig at Wembley Stadium (where, incidentally, they organized the crowd to perform one in the dark, holding up the lights *Surrender to* of their cell phones). I remember one sweeping through *the crowd* the crowd at a football match in London. I liked the way it meant abandoning any sense of my individuality and succumbing to the collective will of the crowd. I suspect this is part of their appeal. Of course, we all consider the prospect of losing our uniqueness in any permanent sense equally horrific (which suggests we might be deluding ourselves about how much we have in the first place). Maybe this fear is why there's a pleasure in relinquishing it–like when a child begs to be chased around the garden by you as a pretend monster because the prospect of a real one is too scary to bear?

Entertaining as it is to watch a wave, and exhilarating to ride one, it's something quite other to actually *be* a wave. You become the medium through which it travels. You are a small part of its collective energy.

~

The giant honeybees of southern and southeastern Asia, *Apis dorsata*, don't build their hives in enclosed spaces like European honeybees. Instead, they suspend their honeycomb out in the open, attached to a high branch or an overhanging rock, and cluster around it as a huge, buzzing ball.[2]

Such a nest is among the most impressive sights in the insect world, in part because of its epic scale–with the 3 by 5ft honeycomb swathed in a protective armor of more than 50,000 wiggling bodies–but also due to the stadium-style waves that regularly sweep across its surface. These are produced by the bees on the outside of the cluster performing what can best be described as highly coordinated apian mooning. Such behavior is known as "shimmering." The insects flick their bottoms in the air in waves.

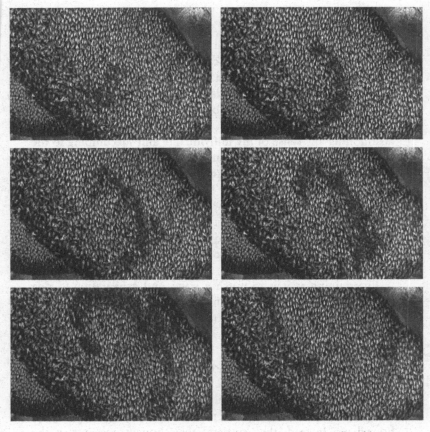

"Bee shimmering"—the most impressive mooning in the natural world.

Since the undersides of their abdomens are darker than their yellow backs, the appearance is of black swathes moving across the surfaces of the colony. These waves tend to spread from a single point in an expanding spiral pattern.

The reason the bees perform this cheeky display? The maneuver seems to have evolved to scare off predatory wasps, such as bee-hawking hornets, which will attack the nest to try to reach the delicious pupae and honey within. Since the bees would die if they deployed their stings, they have developed a less suicidal defense strategy. The wave of shimmering bee bottoms begins at the point on the surface of the hive where the hornet threatens to attack. The overall effect is of something rearing up against the hornet that is far larger than the individual bees.[3] But the tactic is ineffective against larger predators. For when birds attack the hive, the giant honeybees fall back on the old kamikaze stinging missions with such determination that they have earned the reputation as the most aggressive species of honeybee in the world.

Bees vs. wasps

You'd think that the hornets would eventually realize that the wave was just an illusion, but that would be to overestimate their intelligence. Acting as individuals, the wasps are not smart enough to work out the trick of the crowd. It is a classic case of group intelligence outwitting a lone predator, and one that is mediated by a wave that travels through a population.

Waves passing through a group of individuals can also increase the chances of survival when food is scarce. At least, they can for a certain rather remarkable type of amoeba called *Dictyostelium discoideum*.

This is a microorganism that lives in the soil of deciduous woodlands and feeds on the bacteria among the decaying leaves. It is described as a "social amoeba" because of the way the individual organisms get together when times are hard. Normally, when there are plenty of bacteria around for them to eat, they keep to themselves—feeding, replicating, and hanging out as individuals. In this solitary mode, they behave no differently from normal, antisocial amoebas—they don't stop to chat, they never call each other, they just mind their own business. Only when the supply of bacteria runs out and the microscopic organisms are faced with starvation do they finally start to get to know each other.

First you blank me, now you're my best friend?

In fact, they go from one extreme to the other. Some of the cells release a chemical signal, which has a magnetic effect on the neighboring cells; it causes them to move toward the source and, at the same time, release some of the chemical themselves. In this way, the message to gather passes through the *Dictyostelium* population in the form of waves of chemical signal and of gathering movement. When a "lawn" of these amoebas, starved of food in a Petri dish in a lab, is filmed in time lapse these waves of movement appear as dark bands that spread in spirals over the population. Though the individual cells are far too small to be seen by the naked eye (they're only about 10 thousandths of a millimeter long), the surface of the lawn darkens when the cells are moving, on account of them stretching out to do so. The appearance of these tiny rippling bands of light and dark is quite beautiful, but what happens when the *Dictyostelium* get together is less so.

Within a matter of hours, after twenty to thirty waves, the amoebas have congregated into lumps of up to 100,000 cells, called *From amoebas* "slugs." Each slug looks like a tiny blob of Vaseline, about *to slugs* a tenth of an inch long. It bears no relation to the actual, larger slugs you see in the garden, which are individual animals from the mollusk family. Nevertheless, like their namesakes, these blobby collections of amoebas secrete a slimy substance that coats and protects them as a whole—and helps them to move as one.

For the wave-like coordination of the amoebas doesn't stop once they have gathered together. Similar waves of chemical release and movement seem to travel through the population within the slug, coordinating them to propel the slug along as if it were now a single organism. The amoebas in the main body of the slug move forward and backward to form longitudinal pulses that inch it along. Those amoebas at the slug's front tip, which is lifted off the ground, move in a spiral wave, rather like a corkscrew, to ensure that the pulses set off down the slug's length in a coordinated fashion.[4]

Like some 100,000-strong pantomime horse, this slug slithers off in search of a better life, leaving a tiny trail of slime behind it. When it finds somewhere lighter and warmer, it proceeds to rise like a fungus from the ground and form a fruiting body at its tip. A fifth of the population—the ones that drew the short straws—now give their lives for the greater good. Having lifted the rest of the *Dictyostelium* up and up to a towering 1/8in into the sky, they die. The lucky ones at the top change into spores, to be dispersed in

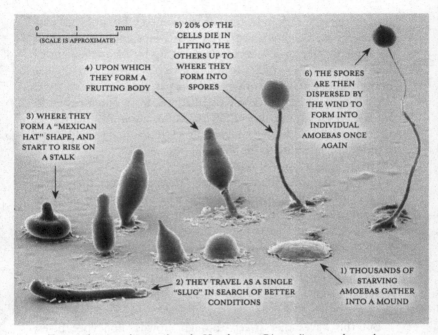

0   1   2mm
(SCALE IS APPROXIMATE)

5) 20% OF THE CELLS DIE IN LIFTING THE OTHERS UP TO WHERE THEY FORM INTO SPORES

4) UPON WHICH THEY FORM A FRUITING BODY

6) THE SPORES ARE THEN DISPERSED BY THE WIND TO FORM INTO INDIVIDUAL AMOEBAS ONCE AGAIN

3) WHERE THEY FORM A "MEXICAN HAT" SHAPE, AND START TO RISE ON A STALK

1) THOUSANDS OF STARVING AMOEBAS GATHER INTO A MOUND

2) THEY TRAVEL AS A SINGLE "SLUG" IN SEARCH OF BETTER CONDITIONS

Teamwork, on a microscopic scale. How hungry *Dictyostelium* amoebas gather, using waves of chemical signals, to form into a slug, which slithers off as they search for a better life.

the wind. When these find themselves in the presence of yummy bacteria once again, they germinate to form more amoebas. And so, as antisocial as ever, the amoebas set about feeding and multiplying once more. They pass each other with barely a nod or hello—that is, until the food runs out again, the waves return and the whole cycle continues.

∿

In most "normal" mechanical waves, energy travels along with the wave shape, while the medium it passes through, be that water, air, rock or whatever, remains roughly where it started. But the waves that sweep through stadiums or across the surface of giant honeybee hives and populations of social amoebas are distinctly different.

In these cases, each individual expends energy, but does not pass this energy onto its neighbor. The sports fans and the bees and the amoebas aren't passive objects swept up by the movement of their neighbors, as particles of water are when swept into circular motion along with all the others with the passage of an ocean wave. No, what passes through these populations, from one member to the next, is something quite different: information. All that travels through the crowd is how and when each individual should move in order to contribute to the wave.

This is quite a contentious idea.

It is contentious simply because "normal" waves themselves transfer information. After all, this is why we've grown sensitive to them in the first place. Thus the light waves we see carry information about the objects they bounce off, and the sound waves we hear tell us about who or what produced them. This is also why we use waves for communication: radio waves, for instance, to carry the signal of a program from the transmitter to your radio. Nevertheless, these are all waves of *energy* traveling from one place to another.

But the waves that sweep through crowds, honeybee hives, and blobs of social amoebas consist of nothing but information. The only thing that travels as a wave from one part of the population to the next is the signal for each individual to do something.

Ask a physicist and he will tell you that stadium waves are not *real* waves because they aren't the movement of energy through a medium, so much as a medium using energy to coordinate itself in some way. But we still see them as waves. This seems as good a reason as any for them to be bona fide waves, along with the wiggling bee bottoms and the congregating amoebas. Who cares if they don't obey the same physical laws as the others? They are cut from a different cloth.

The eggheads may disagree.

But I say 50,000 apian moonings to them.

_C

If you are planning to start a stadium wave, you'll need the help of some friends. Twenty-four, in fact, because you and your pal jumping up with your hands in the air during a Jets game will achieve precisely zilch.

We know this thanks to the studies of Professor Tamás Vicsek, of the Eötvös Loránd University in Budapest.[5, 6] He and his colleagues found that in order to spark a wave in a stadium, you need no fewer than twenty-five people in on the plan, and they all need to coordinate themselves to jump up at once. Whether the twenty-five have to be drunk, with their shirts off and chanting lewdly was not addressed in the research.

"Some people have come to me and said that twenty-five is a wrong figure," Vicsek explained when I asked him about his work. "They've said that they managed to start a wave with just four friends. But it turns out that five people can't start the wave the first time they jump up. It takes them three or four tries, by which time the immediate neighborhood has joined in, making up the two dozen or more needed to get the wave started." *Five is not a critical mass*

Professor Vicsek became interested in stadium waves not because he is an ardent sports fan but as a consequence of research he'd been doing into other crowd behaviors. He had previously worked on a study into how audiences manage to synchronize their applause into coordinated claps when they want to call an act back onstage. Another study looked at the way panic spreads through a large group when something goes wrong. It was at a sports event, while Vicsek was being interviewed for TV about this research, that a wave happened to sweep through the crowd and spark his curiosity.

Vicsek studied how stadium waves begin and propagate by analyzing videos from sporting events. He devised a simplified computer model of the crowd to mimic its behavior. This consisted of an array of virtual people, each a very simplified representation of a crowd member. These could be in one of only three possible states: "excitable," when they are sitting and ready for a wave; "active," when they are standing and waving; and "passive," when they've just waved and won't get up again for a bit. I pointed

"OK, everyone . . . Wait, are we going left or right?"
The 1986 Mexican World Cup: when stadium waves hit the big-time.

out to Vicsek that, for some football crowds, this may not be a simplification.

Even though the "people" in this model were crude, Vicsek and his colleagues could make them accurately mimic the stadium-wave behavior of real crowds. By fiddling with the settings, they could also demonstrate that the speed with which a wave travels around a stadium is determined by the crowd members' reaction times. In reality, it moves fast—at about 27mph.

Experienced crowds are able to introduce a degree of sophistication into their waves. "The freshmen in the University of Notre Dame are very skilled at starting them," said Vicsek. "Learning how to get them started is part of the culture there. They can set a wave moving in whichever direction they choose. They can even make them go both ways at the same time. For that, you need many dozens of people, and you have to really know what you are doing."

Among less proficient wavers, however, it is far from clear what factors cause the wave to travel one way around the stadium rather than the other. Were people merely jumping up because their neighbors are, you might expect the hand waving to spread out from the starting group in an expanding circle, much like the ripples from a pebble in a pond. The wave would thereby travel both ways around the stadium at once—which is the very trick that the fancy Notre Dame crowds achieve only with practice. But Vicsek found that an overwhelming number of waves went only in one direction or the other. In fact, in a second study,[7] a bias emerged: "We found that the proportion of clockwise to counter-clockwise is something like 60:40."

This one included an online survey of stadium-wave participants. Bizarrely, of the seventy-five people who participated in this, all the wavers in Europe remembered their waves traveling clockwise around the stadiums, while 70 percent of those from Australia reported them going counterclockwise. It all seems rather reminiscent of that old myth of water spinning opposite ways down the drain in the Northern and Southern Hemispheres.* However, with stadium waves it seems there may just be some anecdotal truth to a hemispherical difference.

This seemed so unlikely that I decided I had to conduct a study of my own. Not the most robust study, I admit, but I did watch 94 different You Tube videos of stadium waves. (Looking back at it, this was clearly a displacement activity designed to avoid getting on with writing this chapter.) Sixty-nine of them were of waves traveling around stadiums within the Northern Hemisphere. Of these, I counted 40 going clockwise and 29 counterclockwise, a ratio of 58:42 in favor of clockwise. The other 25 videos were of waves at games in the Southern Hemisphere. Of these, 10 waves

---

* The effect of the earth's spin, known as the Coriolis effect, is significant enough to make storm systems rotate in different directions, but it's far too weak to have any significant effect on bathwater, compared with all the other random movements in the tub.

went clockwise and 15 counterclockwise, a ratio of 40:60 in favor of counterclockwise.

I asked a professional statistician if these results were significant. She told me that I can be 96.6 percent certain that the probability of a wave going one way rather than the other is *different* between the Northern and Southern Hemispheres. There is a high probability of a hemispherical difference. It certainly *looks* like waves are more likely to go clockwise in the north and anticlockwise in the south, but I didn't watch enough videos to be able to say this with the 95 percent certainty that is considered significant. That sounds like nit-picking to me. Clearly, this is robust evidence that stadium waves are more likely to go clockwise in the north, and anticlockwise in the south. I hereby name this the "Stadium Law of Hemispherical Bias."

Case closed.

~

You may be interested to hear that hippopotamuses also produce crowd waves.

I'm sorry to say that these do not involve the animals heaving themselves up onto their hind legs and waving their webbed front *The hippo* feet in a coordinated manner that ripples along the muddy *telegraph* banks of the Zambezi River. Although they aren't sticking their feet in the air, they are still producing waves of information—in this case, a form of vocal communication that travels from one pod to the next, dubbed "chain chorusing."

Male hippos will lift their nostrils above the surface of the river and make deafening calls. The fact that it is the males who do the hollering suggests that they might be communicating territorial boundaries. And when Professor William Barklow, of Framingham State College in Massachusetts, studied this behavior he found that it soon gets the neighbors going, too.

The bellowing calls of the males in one pod would cause the members of another, farther along the river, to rise to the surface and roar as well. This group call would, in turn, set off the next pod, so that a cascade of roars would travel down the length of the river. Studying the animals in Tanzania, Barklow observed that this wave of hippo calls traveled as far as 8 miles down a river, taking four minutes to complete the journey.[8]

Besides the racket above the surface, it seems the hippos produce a loud sound under water, too, since their calls encourage those along the river to surface and join in. Barklow calls this "amphibious communication." By lifting its nostrils out of the water while its mouth, lower jaw and throat remain submerged, a hippo can simultaneously produce over- and underwater noise at the same time.

Since sound waves travel over four times faster through freshwater than through the air, the hippos along the river could in theory hear two sounds: first the underwater call, then the overwater one. (Do they perhaps use the difference in sounds to judge the distance of the calling hippo, just as we judge the distance of a thunderstorm by counting the seconds between the flash of lightning and the rumble of thunder?)

Their calls involve complex arrays of tones, apparently including the same sort of infrasound frequencies that elephants use for their own long-distance communication. It seems that the hippo version of a stadium wave is rather more sophisticated than ours. Who knows what the males are saying to each other? Presumably, it contains phrases like "*my* patch," "*my* females," and "back off." At least we can confidently say that it ends with "pass it on."

~

Rome's Piazzale Napoleone, at the edge of the Villa Borghese park, has a terrace overlooking the imposing expanse of Piazza del Popolo below. With its view west, across the River Tiber to where the cupola of St. Peter's Basilica rises from the city's terracotta rooftops, this balcony is a classic spot for a young Roman lad to gaze into the eyes of his love, as together they soak up the sunset.

It also happens to be a great vantage point for a bald Englishman to observe birds of a different feather: the city's flocks of starlings. When I first came across them, one sunny winter evening while I was living in the city, I couldn't at first work out what they were. Coming together as a large aerial flock, the birds collectively formed an amorphous, elastic shape that stretched and shifted and curled in on itself in such a surprising manner that I was mesmerized. From then on, whenever I found myself in the area toward the end of daylight, I would pop up to the balcony and elbow the lovers out of the way to watch the enormous clouds of birds rise and fall, expand and contract, merge and divide. I never tired of the spectacle, and was sad that it lasted no longer than twenty minutes or so.

Flocks of starlings, numbering anything from 200 to 50,000, engage in these aerial displays over Rome on most evenings through the autumn and winter months, upon returning from foraging in the surrounding countryside to roost in the city's treetops. Theirs is the opposite commute from that of the Italians. And the reason for such dramatic flocking behavior? It isn't just safety in numbers. These displays also help the thousands of birds gather and establish their positions in the roosting sites for the night—like an enormous avian game of musical branches, in which everyone gets a perch.

Avian commuters

For some reason, this sight always brought to mind a scene from the classic 1953 Jacques Tati film *Monsieur Hulot's Holiday*. While *en vacance* in a small French seaside town, M. Hulot can't take his eyes off a blob of nougat drooping down from a hook on the side of an ice-cream cart outside his hotel. The gooey fondant sags lower and lower in the warm sun, and looks as if it is about to drop to the ground. Hulot is clearly uncomfortable at the sight of it about to fall, but is unsure whether to intervene and catch it. But the vendor always manages to flop the nougat back over the hook milliseconds before it falls.

I don't know why this image came to mind. After all, unlike the nougat, the starling flocks had a completely weightless quality to them. (The same cannot be said of their droppings, which explain the Romans' hatred for this bird. The flocks may look elegant during their coordinated flights, but they'll think nothing of covering every inch of your beloved Vespa or Smartcar in bird poop should you be unlucky enough to have parked it below a roosting tree. Since the birds now outnumber the citizens of Rome by four to one, this has become a major issue.)

Richard Barnes's *Murmur #21*, showing starlings flocking over Rome.
As the birds perform their coordinated maneuvers, the dark patches, where
they are bunched up, spread through the flocks as waves.

I remember noticing that the waves would flow through the
flocks, from one side to the other, as ripples of bird density. The
dark bands where the birds were momentarily more closely aligned
would flit across the groups, perpendicular to the flight direction.
I remember wondering what these fleeting waves were. Were they a
mechanism to ensure the birds didn't collide with each other, like
some high-speed version of commuters adjusting their positions on
a train to avoid invading each other's personal space? Or did they
mediate some more complex form of inter-bird communication?

The amazing coordination shown by flocking birds has mysti-
fied scientists for a long time. In the 1970s, it was even proposed

that a leader hidden among the flock might be creating an electrostatic field to tell the others when to move.[9] Presumably, it would be saying, "Left, everyone . . . OK, birds, wait a minute . . . now right, right, right!" A study carried out earlier, in the 1930s, proposed that in order for all the birds to change direction at the same moment, they simply used thought transference.[10]

Neither was right, for it turns out that birds sometimes communicate through waves—waves of movement. In a 1984 study, high-speed films of the collective flight of dunlin, coastal wading birds that fly in large flocks, demonstrated that a single bird can initiate a maneuver that spreads through the flock as a "movement wave," helping to coordinate the flight of the whole flock.[11]

But it's not as if they all follow one bird. The films showed that any bird could initiate a change in the group direction, so long as it banked toward the flock, rather than away from it. When a bird turns away from the others, it will be largely ignored. This directional rule may have developed to stop flocks breaking apart and to improve response to attacking birds of prey, which tend to pick off isolated individuals. Indeed, defense against aerial attack seems to be one of the main reasons for these waves of movement. News of a predator's approach can be communicated rapidly through the flock by whichever of the hundreds of birds on the outside notice it first.* When under attack by a peregrine falcon, for instance, starling flocks will contract into a ball and then peel away in a ribbon to distract and confuse the predator.

....

* This rapid transfer of information through a flock is sometimes called the "Trafalgar Effect," after the British tactic, at the battle in 1805, of using flags to communicate from one ship to the next that the Franco-Spanish ships were setting sail from Cádiz. While Nelson's fleet was stationed 50 miles offshore, well out of sight of the coast, it was connected by a line of vessels, each within view of the next, to four frigates close to the port. This meant that the signal of flags announcing that the enemy were leaving the safety of the harbor was passed rapidly down the chain of ships to Nelson aboard HMS *Victory*, which was over the horizon line. The news was communicated farther than the reach of any telescope and in far less time than any craft could have traveled, giving the British a head start to position themselves for the battle. The same principle—though for defense rather than attack—underpins the rapid communication through a flock of starlings that a falcon is approaching. The birds pass the information on with rapid sideways movements, rather than by hoisting flags.

The films showed that at the start of a change in direction, each bird takes around 70 milliseconds (ms) to react to the movement of its immediate neighbors. But once the collective wave motion has started sweeping through the flock, they change direction only some 14ms after their neighbors. Interestingly, the average reaction time of dunlin, when measured in a lab, is only around 38ms. It seems the sight of the wave sweeping through the flock helps the birds react faster than if they were changing direction according to the movements of their immediate neighbors alone. By seeing the wave coming, they are ready to move milliseconds earlier than they would otherwise be, much like a glamorous, sequinned dancer watching the wave of high kicks coming down the chorus line in order to time her own kick with its arrival. For this reason, the theory of how birds coordinate while flying in flocks is sometimes known as the "chorus-line" hypothesis.

High-kicking dunlin

Coordinated starling flights have evolved over millennia, but crowd waves in stadiums are obviously an invention. So when did the first ones appear? How did the first crowd of spectators ever coordinate themselves to make a wave go around the stadium?

As we have seen, a crowd needs to coordinate itself consciously to form the wave. The wave can only be set off around the stadium if there are enough potential participants who know what they are trying to do. How did a crowd ever get itself coordinated enough to produce one before it had become such a celebrated activity?*

According to the official website of the football team of the University of Washington,[12] the very first crowd wave took place in Husky Stadium, Seattle, on October 31, 1981. The home team was playing against Stanford University.

....

* Of course, waves have always naturally spread through crowds. Football chants, rugby songs and the ubiquitous slow handclap are examples of how the behavior of some vociferous individuals can spread through the crowd until most of the stadium are joining in. But these waves happen unintentionally. They are a result of the time it takes information and enthusiasm to spread, infecting one part of the crowd after another.

The prototypical stadium wave was instigated by Robb Weller, a Husky alumnus. During his college years, Weller had been the team's cheerleader. As a matter of historical interest, cheerleading was originally an all-male activity and, luckily for Weller, didn't involve wearing short skirts and pom-poms, just coordinating the crowd to help whip up the excitement. This Stanford fixture was the Homecoming game, held when the university opens its doors to its alumni for a week of festivities. That day, Weller stood on the sidelines along with Bill Bissell, former director of the official Husky Marching Band, microphone in hand, to lead the crowd as they'd done a decade before.

In the third quarter of the game, Weller and Bissell tried to get the home crowd to stand up, starting with the lowest seats and progressing to the highest. This would have been more of a concentric wave than the one we all know and love. Alas, it proved too difficult to coordinate the crowd in this pebble-in-a-pond style of movement. Weller did, however, manage to get the crowd to stand up in a wave that swept around the U-shaped Husky Stadium from one end to the other. According to the team website, "The original Wave saw Husky fans remain standing until a full circle was completed in the stadium."[13]

Weller and Bissell returned to lead the twentieth-anniversary wave at the Huskies–Stanford Homecoming game in 2001.

So that's that then.

At least, I thought it was until I came across this:

MY STATEMENT OF IRREFUTABLE FACT:
I, KRAZY GEORGE, INVENTED THE WAVE.
I ORCHESTRATED IT ON OCTOBER 15, 1981 BEFORE
A NATIONALLY TELEVISED AUDIENCE AND A
SOLD-OUT STADIUM DURING THE AMERICAN
LEAGUE PLAYOFF SERIES BETWEEN THE
OAKLAND A's AND THE NEW YORK YANKEES.

"Krazy" George Henderson is a well-known face among the stadium crowds in his home state of California. A former school-teacher, he now describes himself as a "professional cheerleader." You can't miss him up at the front of the crowd: bald on top with white flowing locks on the sides, banging his signature snare drum and yelling instructions at the crowd.

Now, I'm inclined to dismiss the assertions of anyone who writes in capital letters (which is how Krazy George formats his wave-inventing claims on his website[14]). That said, there might just be something in his claim.

The baseball game to which he refers took place at the Oakland Coliseum stadium a full sixteen days before the Huskies game. As usual, Krazy George was whipping up the crowd in his section of the stadium. "I told everybody to stand up as I sweep my hand. Stand up, yell and sit down." If the wave stopped, he told them they should boo. The first two attempts went around only a few sections of the stadium. When they stopped, George's section booed. "The third time, it went all the way around. Then the whole place stood up and applauded," he remembered with pride.[15]

So where's the proof? Well, as it happens, the Oakland As' 1981 season highlight film shows the wave taking place at the A's vs. Yankees baseball game. That, you'd think, would be the end of the matter.

But when Jon Cudo, who runs a company that books entertainment at sporting events, posted a story on his website, Gameops.com, crediting Krazy George with the invention of The Wave, he received angry posts from Husky fans,

Krazy George Henderson: he may be a professional cheerleader, but did he invent the stadium wave?

*"... YEAH! IT'S LIKE A WAVE OF PEOPLE THAT SWEEPS
AROUND THE STADIUM!"*

accusing him of bias. They pointed out that he had worked with the events booking agency that represents Krazy George. Cudo claimed on his site,

> *I spent a lot of time working on a story years ago regarding the wave and the claims of George and the University of Washington. After hours of time and several conversations with people in the Athletic Department at UW, I was left with this: They see their claim as an urban legend that they are happy to reinforce. However*

> *they stopped short of explaining how Oct 31, 1981 (which is the*
> *date they claim they "invented" the wave) came before Oct 15,*
> *1981 (the day George has video of a wave he started in Oakland).*[16]

But he refused to be drawn any further. "I (literally) had threats of physical harm from UW fans after we posted our story."

Cudo now ignores all emails, posts or letters on the subject. (And so shall I.)

⌒

In *The Mezzanine*, the novelist Nicholson Baker describes in loving detail the minutiae of a lunch break, as experienced by a young office worker named Howie. At one point, Howie pays a visit to the men's room, where he overhears a coworker whistling a florid rendition of "I'm a Yankee Doodle Dandy."

This reminds him of an occasion when, "hopped-up" on the first coffee of the day, he launched into a fulsome whistle of "Wouldn't It Be Loverly?" from *My Fair Lady*, only to realize that he had drowned out his cubicle neighbor's restrained interpreta-   *Musical* tion of some soft-rock classic. The incident had caused   *lavatories* him some embarrassment at the time, but later that day he had been delighted to hear a somewhat embellished version of his tune being whistled by someone over by the photocopier. They must have been listening from one of the other cubicles when he had whistled roughshod over his colleague.

His reminiscences over, Howie finally makes his way back out of the lavatories, striding into the office hallway, only to catch himself whistling "I'm a Yankee Doodle Dandy." Songs can be infectious, it seems.

The present-day champion of evolution theory, Richard Dawkins, invented the term "memes" to refer to "tunes, ideas, catch-phrases, clothes fashions, ways of making pots or of building arches" that have an infectious quality to them. Just as genes propagate by leaping from body to body via sperm or eggs, "so memes propagate themselves in the meme pool by leaping from brain to brain via a process which, in the broad sense, can be called imitation."[17]

Nicholson Baker's whistling memes are little waves of information that spread through an office. But, unlike stadium waves, these ripples of imitation happen subconsciously. They pass through uninvited.

In Fritz Leiber's irreverent sci-fi short story from 1958, "Rump-Titty-Titty-Tum-Tah-Tee," a group of ambitious intellectuals gathers at the studio of an avant-garde "accidental artist" to watch him create one of his celebrated splatter paintings. A famous jazz drummer, Tally B. Washington, is among them, and sits tapping out crazy rhythms on a hollow African log as the artist positions himself, loaded paintbrush in hand, over the huge blank canvas. Tally happens to tap out the rhythm of the story's title on the log just as the artist launches a volley of paint in the air. The paint happens to land on the canvas with the *exact* same succession of beats: *rump-titty-titty-tum-TAH-tee!* The intellectuals, stunned by this coincidence, become obsessed by the catchy rhythm. They subsequently can't help but disseminate it through their various creative fields.

So infectious is the rhythm, in fact, that it spreads virally across the world. A new style of music, based upon these beats, then becomes an international craze. People start carrying around Blotto Cards, which reproduce the shape of the paint splatter. *Rump-titty-titty-tum-TAH-tee!* soon threatens to bring humanity to its knees. "Can't get it out of our minds. Can't get it out of our muscles," despairs a psychiatrist member of the group when the intellectuals gather again a few weeks later. They are all held in "psychosomatic bondage."

The only chance they have of stopping the musical pandemic is to conduct a seance. So doing, they discover that a distant ancestor of Tally's, the jazz drummer, had been an evil witch doctor and must have released the rhythm maliciously by possessing his descendant. Only by contacting the spirit of the witch doctor are they then able to halt the rampaging rhythm's progress.

This meme may not be quite what Richard Dawkins had in mind.

Fritz Leiber's infectious rhythm may be an amusing example of an idea spreading like a wave—or perhaps, more accurately, a tsunami—of information. But it might not be quite as fanciful as it sounds. Take the oscillating fortunes of the financial markets. Are the rising and falling values of the NASDAQ, the Dow, and the Hang Seng indicative of some sort of wave? The Elliott Wave Principle of economics would seem to suggest so.

Stock-market waves

Ralph Elliott, an American accountant, devised the principle in the 1930s, having lost his job and much of his fortune in the Wall Street Crash of 1929. In essence, Elliott held that the bull and bear phases of stock markets develop in a series of waves. And just as the ripples on the surface of the ocean can be superimposed upon larger waves, which in turn are superimposed upon tides, these financial oscillations occur concurrently at different time-scales. Whether "minor," "intermediate," "primary," or "grand" cycles, each, he claimed, showed a pattern of five waves of growth, followed by three corrective waves of decline.

To be honest, I don't know whether the peaks and troughs of the stock markets qualify as real waves of information. Perhaps the periods of boom and bust just *look* like waves when you plot them on a graph against time. But even if the wave is just a metaphor, it is a compelling one. "Waves" of confidence that dictate the fortunes of the global markets are essentially the sum of our states of mind, the aggregation of all our individual senses of financial security, which inform our everyday decisions on borrowing, saving and investing.

Which must then mean, logically, that the traders and fund managers are the surfers. They are always trying to catch the waves of confidence at the right moment, to ride them relentlessly and bail out just before it is too late and everything comes crashing to the ground.

And what about flu? Passing from one human host to another, it can cross countries as an epidemic and continents as a pandemic. Each virus contains a genetic code, which it spreads as it multiplies within the cells of the infected hosts, and between them by means of sneezes and contact. The flu pandemic of 1968 spread from Hong Kong right across the world in a matter of months. The mortality rate was relatively low, around a million

Flu-pandemic waves

deaths worldwide. The 1918 Spanish flu, by contrast, killed more than 40 million people in just a year, and was estimated to have infected nearly a third of the world's population.[18] But both pale in comparison to the Black Plague, a bacterial rather than viral pandemic, which is thought to have killed around 60 percent of all Europeans living in the middle of the fourteenth century.[19]

In 2009, the world waited to see how deadly would become the H1N1 swine flu pandemic that had begun in April in the state of Veracruz, Mexico, and spread in a matter of months to most countries around the world. By the end of the year, it had claimed almost 13,000 lives in 130 countries,[20] but what threatened to become an unstoppable wave of death now turns out to have been more a wave of panic—and of profits for Roche, the Swiss pharmaceutical giant, which was forecasting toward the end of 2009 that the year's worldwide sales of the antiviral drug Tamiflu would reach $2.65 billion.[21]

Even if the number of deaths from the initial outbreak of the virus was, thankfully, far lower than previous flu pandemics, the virus still swept like an enormous wave through the world population. The Centers for Disease Control and Prevention estimated that the number of people contracting the virus in the United States alone was anywhere between 39 and 80 million.[22] This particular Mexican wave was both uninvited and potentially uncontrollable. Its geographical spread was not straightforward like a wave that sweeps through a stadium. Rather, it traveled along networks of mobility, dictated largely by flight paths. Nevertheless, it was a wave of genetic information, for which the world human population is the medium. As people recover from the virus, it spreads like a regular wave. Thankfully, it seems to have failed to turn into the medical equivalent of a shock wave: when the human-population medium through which it travels is *permanently* affected by the wave. By this, I mean people not returning to normal health, but dying.

The "credit crunch" that came to a head in 2009 was like a financial shock wave. The crest of economic confidence had become so top heavy that the system was perilously unstable. The pattern undulations, whether gentle or dramatic, broke down entirely, with markets crashing over themselves in a maelstrom of economic turbulence and dashed fortunes. There were the usual reversible trends of a financial downturn—the plummeting shares, plunging property prices and mounting unemployment—but also irreversible and widespread damage, as banks across the world had to be shored up by governments, and major financial institutions collapsed entirely.

All this chaos and disorder seems a far cry from good old stadium waves. Those innocent waves of information are, after all, extraordinary expressions of collective control. The pleasure of a crowd wave is seeing everyone participate in the same pointless *Making waves* activity that is both synchronized and spontaneous. If *for the* we can organize ourselves enough to make these waves *greater good* sweep around stadiums, perhaps we can similarly coordinate our efforts to tackle modern-day pandemics—whether medical, financial, or other—before they develop the irreversible destruction of shock waves.

The Seventh Wave

WHICH EBBS AND FLOWS

Which are the largest waves on the sea? A friend asked me this when we were having a drink one September evening, as if it were a question in a pub quiz. For a moment, I imagined the two of us competing against the other tables around the room as this very question was being called out by the quizmaster at the bar. My willingness to picture this scene was not because I have a particular fondness for pub quizzes. In fact, I absolutely hate them for I invariably lose whenever I play. The reason I liked the idea of being on a team when this one came up was because it is a gift of a question: one to which most people think they know the answer, but will in fact get wrong.

Zero points to the table over by the bar for incorrectly guessing that the biggest waves are the huge breakers that come crashing ashore in the wake of hurricanes and other violent weather systems.

Waves of Hurricane Carol wreaking havoc in Old Lyme, Connecticut, 1954.

Same for the smug regulars over by the fire, whispering and scribbling down the 60–100ft-high rogue waves that have been *Wrong again!* reported for centuries by mariners as appearing unexpectedly amid violent sea storms and forming immense walls of water that are out of all proportion with the other storm waves.*

. . . .

* Satellite imagery and wave measurements from oil platforms have confirmed that these monster waves exist. It is now thought they might have played a part in the quarter of large-vessel shipwrecks that are recorded as "unexplained." Indeed, the crews of many of the two hundred or so supercarriers lost in high seas between 1981 and 2000 have reported individual, or groups of, rogue waves as playing a part in the wrecks. (See Rosenthal, W., and Lehner, S., "Rogue Waves: Results of the MaxWave Project," *J. Offshore Mech. Arct. Eng.*, vol. 130, issue 2, 2008.)

A rogue wave looms astern of a merchant ship in heavy seas in the Bay of Biscay,
off the coast of France, *c.* 1940.

Even that know-it-all table, sitting silently with steely expressions, would have got it wrong. The biggest waves are not *Still wrong* tsunamis, as they'd have believed, like the one caused by an earthquake on December 26, 2004, that led to an estimated 230,000 deaths on the shores around the Indian Ocean.

The Indian Ocean tsunami arriving at Ao Nang, Thailand, 2004.

The correct answer—and this is where my friend and I would have cleaned up—is a much, much larger wave than any of these. This one, in fact:

That's wiped the smiles off their faces.

It may not look like much, and it doesn't have a very impressive height—from crest to trough, just a couple of feet—but it is absolutely enormous by that other measure of wave size, namely the length from one crest to the next. This can be many hundreds of miles. It is a "tide wave."

"Tide wave" should not be confused with "tid*al* wave," a term sometimes used to refer to a tsunami. In fact, tsunamis have nothing to do with the tides: they are caused by sudden events, such as the *"Tidal waves"* seabed shifting violently in an earthquake, causing a huge *now banned* disturbance in the water level that travels across the ocean. This inappropriate and confusing term will henceforth be banned from this book, especially since the consistent use of "tsunami" in the blanket media coverage of the horrific events of December 2004 has ensured that "tidal wave" has finally been swept from common usage.

Tide waves are produced by the gravitational pull of the sun and the moon. The effect of these two celestial bodies is to make the earth's oceans bulge slightly on the nearer and farther sides of the globe. While these bulges might not sound much like waves to you, remember that the earth's spin causes them to sweep across and around its oceans as they try to stay aligned with the positions of the sun and the moon.

You might think of this gathering of water as being like the movement of iron filings in the presence of a magnet. A flat layer of filings spread out on a piece of paper bears little similarity to an ocean of water over a spherical globe, but the movement of the filings rising in a bulge when the magnet is suspended an inch or so above them brings to mind the tidal bulges of the oceans. Just as a ripple of raised filings will travel across the layer as you slide the paper over the table below the magnet, so a bulge will travel across the oceans as the earth spins. That said, the means by which the gravitational forces of the sun and the moon move the water is far more complex than the magnet's effect on the filings. So is the way that the oceans respond to these forces. The iron-filings-and-a-magnet notion is clearly a huge simplification.

But tides certainly are waves, whose peaks and troughs account for the regular rise and fall of the water level at our shores. And, crest to crest, they dwarf all the other waves that form on our oceans.

What's that? A drink for the victors? I think I'll have a pint of stout, thanks very much.

But can it really be right to call tides waves? "Normal" ocean waves—the ones caused by the wind—are waves, rather than currents, because they travel across the water while the water itself remains pretty much in the same place. Tides, on the other hand, are currents of water, aren't they?

Anyone who has watched the tide coming in at Morecambe Bay, on the Lancashire coast, can be in little doubt. Sometimes described as England's "wet Sahara," the sand flats here cover an area of 120 square miles. When the tide arrives, it sweeps over an expanse of gullies, inlets and sandbanks at a speed of 10mph. This is an extremely dangerous area to be in without an experienced guide and a good tide table, and is made all the more perilous by the quicksand that forms as the incoming water fills the hidden holes, known as melgraves, gouged out beneath the sand by previous tides.

The perils of the incoming tide were tragically demonstrated in February 2004, when twenty-one illegal immigrant workers from China were drowned while harvesting cockles in Morecambe Bay. As a storm developed off the coast, the incoming evening tide outpaced them over the sinking sands, cutting off their route to the shore, and safety. Apparently, local cockle pickers had tried to warn them against setting out across the sands at that time of the day by pointing at their watches, but the immigrants' lack of English and the culture of distrust between rival pickers meant that they had not heeded the warning. When they became stranded, one of the workers used a mobile phone to call the emergency services, but the operator was unable to understand where they were. She rang the number back, but the only words she could hear being screamed down the phone were "sinking water," and "the sound of wind and water and other foreign voices shouting and crying out in the background."[1] It later transpired that the workers had been told the wrong tide times by their gangmaster.

Treacherous sands

But the tidal movement of water is nowhere more dramatic and powerful than when a "spring" tide—one forming around a full or new moon, when the "tidal range" from low to high water is greatest— has to pass through a narrow channel.* Such is the case in the Hebrides, where the incoming tide is constricted as it flows through a strait called the Gulf of Corryvreckan, between the islands of Jura and Scarba off the coast of Scotland. On the flood of a spring tide, the current passing through this channel can reach 8.5 knots, or 10mph. As it interacts with the uneven seabed, the flow forms powerful eddies, standing waves and upthrusts of water. And where it passes over a hole within which the bed sinks

Treacherous waters

....

* The term "spring" refers to the verb, as in to spring up, rather than to the season. A spring tide is when the tidal range is large, and can happen at any time of the year.

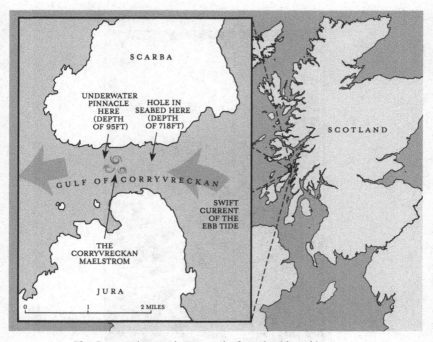

The Corryvreckan maelstrom results from the tide rushing over
a very uneven sea floor.

to a depth of 718ft, followed by a pinnacle of rock that rises to
95ft, the current forms a treacherous whirlpool, known as the
Corryvreckan. This name derives from the Gaelic word *coirebhreacain*,
meaning "cauldron of the speckled sea."

The Corryvreckan appeared in the climax of a 1945 film by
British director Michael Powell. The film's heroine, on her way
to marry a wealthy industrialist on the fictional Hebridean island
of Kiloran, is stranded en route, on Mull, where she begins to fall
in love with a handsome naval officer who is also trying to return
to Kiloran. Torn between her burgeoning feelings for him and her
plans to marry into money, she insists on making the journey in
spite of rough seas and stormy weather. Needless to say, the naval
officer insists on accompanying her, and becomes a hero when the
engine is flooded in the gale as they are sucked into the swirling

Ichiryusai Hiroshige's *Rough Seas at the Whirlpools of Awa* (1853–56) depicts the whirlpools caused by the tide rushing through the Naruto Strait, Japan.

waters of the Corryvreckan. This classic film is called *I Know Where I'm Going!*—a phrase you'd certainly hope to hear from the boat's helm were you being ferried down the strait amid the rushing torrent of a spring tide.

Such tidal currents rush through many constricting channels around the world. The Naruto Whirlpools of Japan, for instance, are produced by the tide surging in between the islands of Awaji and Shikoku, through a channel that connects the Pacific Ocean and the Seto Inland Sea. Norway, however, seems to have more than its fair share of these turbulent currents. The violent swirling eddies of the Lofoten Maelstrom, between Lofoten Point and the island of Værøy, just off the northern coast, within the Arctic Circle, flows as fast as 10–11 knots, or around 12mph. This area was the inspiration for the dramatic whirlpool in Jules Verne's *Twenty Thousand Leagues Under the Sea*. But the fastest tidal current in the world is probably the one that forms the maelstroms beneath the bridge at Saltstraumen, southeast of the Norwegian town of Bodø. Here the current reaches speeds of 22 knots, or 25mph.

Tidal currents are distinguished from normal ocean currents—those driven by the rising of warmer waters and sinking of colder ones and the planet's most consistent wind patterns—by the fact that they change back and forth on a daily basis. Along most coasts, such as those around the Atlantic, the tides are "semidiurnal"—that is, ebbing and flowing twice each day—while on others, such as those around the Gulf of Mexico and the South China Sea, there is just a single rise and fall, known as a "diurnal" tide.* Even if they do change direction regularly, they are still currents, so how can a tide also be a wave? Since they are so broad and shallow, we never notice the passing of tide waves out at sea. We only ever notice them when they interact with the land. Yet when you compare tides to normal, wind-generated ocean waves at a gently sloping shore, they do actually begin to look similar.

We know that normal ocean waves are transformed as they approach the shore: from deep-water to shallow-water waves. As the depth of the water decreases the wave can no longer move in the circular paths that it follows when it is out at sea. Its movement is constricted into increasingly flat orbits by the rising seabed, until only the forward and backward component of the water's movement remains—the familiar washing of the waves up and down the beach. They've changed from deep- to shallow-water waves once the water depth is about one-twentieth of the wavelength.

....

* And some regions, such as the Pacific coast of North America, experience a "mixed" tide, with two high tides a day (like the semidiurnal) but at considerably different levels from each other—one higher high tide and one lower high tide. The reason for all the differences in tide frequencies around the world is partly due to a coast's latitude, as explained in this diagram, which will, I'm sure, greatly clarify things for readers.

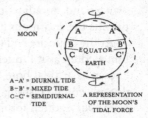

MOON

A–A' = DIURNAL TIDE
B–B' = MIXED TIDE
C–C' = SEMIDIURNAL TIDE

A REPRESENTATION OF THE MOON'S TIDAL FORCE

Tides, on the other hand, are always shallow-water waves. Even the deepest parts of the deepest oceans, which lie 7 miles or so below the sea's surface, are far less than one-twentieth of the wavelengths of tide waves, which, measured from one tide-wave crest to another, can be hundreds of miles. The fact that we see the tide waves arrive and recede as currents is a matter of scale. Compare yourself watching the tide coming in to a hermit crab watching a normal ocean wave wash forward and back over the sand in the shallows at the water's edge. As far as the crab is concerned, the arriving shallow-water wave is simply the wash and backwash of water over the sand. We experience the same sort of effect from the arrival of the shallow-water tide waves. Tide waves and ocean waves differ not only in their scale; there is also the length of time between successive crests and the regularity of their arrivals. Ocean waves, traveling freely across the sea and combining with others, can arrive in quite random patterns, while tide waves, being "forced" by the metronomic movement of the celestial bodies, are far more predictable.

Although most of us are familiar with the ocean tides, few of us are aware that similar gravity-induced waves also travel through the solid earth itself. These "earth tides" are formed as the slightly elastic rock of its crust is distorted by the same combined gravitational tug

Earth tides

of the sun and moon. The earth's tidal bulge leads to a rise in the ground level of up to half a meter at the spring tides. Like those on the oceans, this earth tide sweeps around as the planet spins, trying to stay aligned with the gravitational forces. Since it is not a liquid, the crust experiences none of the currents caused by the ocean tides. Being so subtle, the earth tide is hard to measure, but it is significant enough to require physicists at CERN, the European Organization for Nuclear Research at Geneva in Switzerland, to adjust their calibrations to account for the way it distorts the enormous loop around which their particles race.

At the opposite end of the subtlety scale is when an ocean tide is bunched up as it funnels into a river estuary to form a steep front, known as a "tidal bore." Unlike a normal ocean wave, a tidal bore doesn't come crashing ashore. Rather, it progresses up a winding river course, often for many miles. Sometimes it appears as a series of smooth undulations, which might be led by a bubbling line of white water or even a plunging breaker that curls over itself in the form of a tube.

Tidal bores have been recorded on at least sixty-seven rivers around the world.[2] They are known to form on thirteen British rivers, though some of these bores are so puny and unspec- *Our most* tacular when they arrive that they rather live up to their *exciting bore* name. The UK's most exciting bore forms on the lower reaches of the River Severn. At the mouth of this river the tidal range—which is the difference between water levels at low and high tide—can be as much as 45ft,* producing a bore with a face of just over two and a half meters on the spring tides at the full or new moons nearest to the spring and autumn equinoxes.

Because it just keeps on going at the front of the flood tide, you'd think that the Severn bore would be a wave worth surfing. It promises a ride that would last far longer than a normal wave rolling up the beach. This, at least, must have been the hope of Lieutenant Colonel John Churchill when, in the 1950s, he became the first person brave enough to try surfing a tidal bore.

Known as "Mad Jack" because of his exploits in World War II while leading a commando force, John Churchill was also a keen archer who had represented Great Britain in the 1939 World Championships. He often went into battle carrying his longbow, with which he would loose arrows at the enemy. Though an Englishman, he was obsessed with the bagpipes, and played "The March of the Cameron Men" on the bows of the landing craft as his company made an assault on Vågsøy in Norway. In keeping with the Scottish theme, he would lead his men ashore roaring loudly and holding aloft a broad-hilted claymore without which, he claimed, an officer was "improperly dressed."

Even in peacetime Churchill liked to make an impression. He was fond, for instance, of startling fellow London commuters by throwing open the window of the train and tossing his briefcase

---

* This range is the second highest in the world, beaten only by the 50ft maximum range at the Bay of Fundy, on the Atlantic coast of Canada, between the provinces of New Brunswick and Nova Scotia.

That's Mad Jack Churchill at the front, carrying his trusty Scottish broadsword, during a training exercise with his commandos in the 1940s.

into the darkness. (Little did they know that the old scoundrel had carefully timed the lob so that it would land in his garden, and he wouldn't have to carry it home from the station.)

After the war, Churchill had taught at an RAAF base on the south coast of Australia, where he'd developed a passion for surfing. In the autumn of 1954, back in England, he paid a visit to Frank Rowbotham, the district engineer for the Severn River Board. After swearing Rowbotham to secrecy, Churchill confided that he was determined to surf the Severn bore, and wanted some advice; only too glad to help, Rowbotham explained the behavior and timings of the bore. And so it was that, at 10:30 a.m. on July 21, 1955, the forty-eight-year-old adventurer paddled out from the bank at Stonebench to meet the incoming bore, and clambered onto his board just as the wave arrived.[3]

It wasn't exactly a marathon ride. Churchill stayed on the front of the bore for a couple of minutes, but fell in as it passed through the shallower water, less than half a mile upriver, where a ledge of rock extends across most of the riverbed. The change in depth caused the bore to tumble over and break, as it does at several points along the way, and the unexpected turbulence caused Churchill to lose his footing. He had hoped to ride the bore for many miles, so after struggling ashore, he determined to return the following year.

Though he never did return, "Mad Jack" Churchill had started something. In his wake, modern-day bore surfers have learned to perfect the technique of staying on the front of the surging tide. In fact, locals surfing the Severn bore for many years held the record for the longest continuous surf in the world.

∽

Ever since I was six months old I've been deaf in my right ear. It has never really caused me that many problems, other than rendering impossible dinner-party chitchat with the person sitting on my right. However, being half-deaf has made me useless at determining the direction a sound is coming from. So if I'm out walking and I hear a bird call, I'm incapable of determining where it is coming from. And when I lose my mobile phone and dial the number to make it ring, I stumble around the room trying to work out whether the ringing is getting louder or quieter, depending on my proximity to it.

But I've at last found an advantage to being deaf in one ear. It's given me a way of understanding a rather puzzling thing about the tides: that they should be influenced by the moon more than by the sun. The reason this is puzzling is because *Partial deafness, a silver lining* the gravitational pull of the moon on the earth is much weaker than that of the sun. Boffins have worked it out to be about 178 times weaker,* so it seems a little strange that the position of the moon accounts for around 68 percent of the control on the

* The sun is, on average, about 390 times farther away from the earth than the moon, but its mass is 27 million times as great as the moon. Thanks to Sir Isaac Newton, we know that the gravitational force of a celestial body on a unit of mass is proportional to the mass of the body divided by the square of its distance away. So the gravitational force of the sun upon the earth, compared to that of the moon, is $27 \text{ million}/(390)^2$, or about 178.

tides, compared with the sun's 32 percent. Why should the weaker gravitational pull have the greater tidal effect? I'll try to explain this by means of one ability that I lack: directional hearing.

One of the ways that we—or should I say you?—determine where a sound is coming from, such as that of a mislaid mobile phone, is subconsciously to compare the intensities of the sound waves reaching each ear. The difference in volume at each ear helps you work out the angle of the phone, how far to the left or right it is from the direction you're facing. No difference, and you know the sound's directly in front or behind you. A big difference, and you know it's off to one side—whichever one is the louder.

We all know that it is easier to hear the direction of a nearby sound than a far-off one, but how many of us know why? The reason is that the difference between the ears is more pronounced when the source of the sound—a ringing phone, for instance—is *Where did I* close than when it is far off. The overall volume of the *put my blinking* ringtone is not important (so long as you can hear the *mobile phone?* sounds clearly), only the *difference* in the volume in each ear. And there is a still greater difference in the intensity of sound at each ear from a nearby phone with the volume turned right down than a distant one up at full volume—in other words, the difference between the ears is more pronounced for a close sound than a distant one, even if the more distant is louder overall.

The reason it's easier to judge the direction when your phone is nearby lies with the fact that sound spreads out in a spherical shape. Were you able to see the sound waves, these variations in the air pressure might look a little like expanding balloons, growing extremely fast from the phone but without bursting. Just as the intensity of a balloon's color fades as it grows, with its pigment being spread over a larger and larger surface area, so the intensity of the sound also fades as its energy is spread over a larger and larger sphere.

And here is the crucial factor in our subconscious calculation of its direction: how much the ringtone's intensity changes over the extra few inches the sound waves travel beyond the ear that is nearer the mobile to reach the ear that is farther away. The change in intensity is a matter of geometry: it depends on the difference

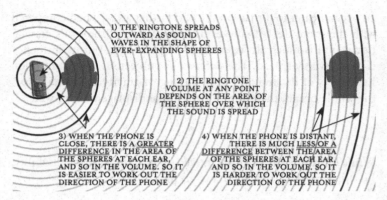

1) THE RINGTONE SPREADS OUTWARD AS SOUND WAVES IN THE SHAPE OF EVER-EXPANDING SPHERES

2) THE RINGTONE VOLUME AT ANY POINT DEPENDS ON THE AREA OF THE SPHERE OVER WHICH THE SOUND IS SPREAD

3) WHEN THE PHONE IS CLOSE, THERE IS A GREATER DIFFERENCE IN THE AREA OF THE SPHERES AT EACH EAR, AND SO IN THE VOLUME. SO IT IS EASIER TO WORK OUT THE DIRECTION OF THE PHONE

4) WHEN THE PHONE IS DISTANT, THERE IS MUCH LESS OF A DIFFERENCE BETWEEN THE AREA OF THE SPHERES AT EACH EAR, AND SO IN THE VOLUME. SO IT IS HARDER TO WORK OUT THE DIRECTION OF THE PHONE

Finding your misplaced phone as its rings depends on hearing the subtle difference in volume at each ear—or, in my case, asking someone else to find it.

between surface areas of the "spheres" of the sound waves at each ear. When the phone is nearby, the spheres are both small, and the few extra inches of diameter of the spheres at one ear and at the other makes a large difference, proportionally, to their surface areas. When the phone is on the other side of the room, the same few inches difference in diameter between the spheres at each ear makes much *less* of a difference, proportionally, to their surface areas.

Now that you've found your phone, you are probably wondering if I am going to get around to explaining why the moon has a greater influence on the ebb and flow of our tides than the sun.

The fixed gravitational field of the moon is not exactly analogous to a sound wave emanating from your phone in an ever-increasing sphere, but you can certainly think of the moon's gravitational field as being spread over a larger and larger sphere with increasing distance. For this reason, its intensity decreases in the same proportions as the intensity of the sound waves decreases. Even though the distances involved are vastly different, both intensities change because they are spread over larger and larger spheres.

Remember that the *difference* in sound at each ear is more marked for a close ringtone than a distant one, even if the volume of the close phone is turned down, and so the sound is *quieter* overall than the distant phone. The same applies with the gravitational forces. The closer moon has a more marked *difference* in pull on the part of an ocean nearer to it than on the part farther away, when compared with the more distant sun—even though overall the sun's gravitational pull is much stronger. It may be stronger, but its influence is more evenly spread. The *variance* in pull causes the ocean tides. Regardless of its strength, were the tug of gravity the same across the whole ocean, there would be no tidal displacement of water.

While I'm aware that we have moved on from my hearing disability, I do find myself strangely drawn to this:

"Has anyone seen my cell phone?"

Why? Not because I think an ear trumpet is a particularly stylish look, but because the funnel of an ear trumpet demonstrates rather well how an estuary can cause the front of the flood tide to form into a bore.

The best bores occur where the coastline around a river mouth is in the shape of a large V so that the space for the tide becomes more and more constricted as it flows inland. The levels of the seabed and riverbed should also rise steadily so that the water becomes increasingly shallow as the tide progresses, further constricting its flow. And last, but emphatically not least, the area needs to experience a sizeable tidal range.

No wonder the River Severn forms a good bore. The Bristol Channel feeds into the Severn estuary like an enormous funnel, with the southern coast of Ireland further adding to the constricting shape. And the funneling continues along the river. The channel is about 5 miles wide near the port of Avonmouth, narrowing to a little under 1 mile by Sharpness, 15 miles upstream. After another 20 miles, having tapered all the way, it is only about 165ft across. The water depth also has a constricting effect. It decreases steadily from around 330ft south of Ireland to 10 or 14ft (during low tide) around the mouth of the Severn.

An estuarine ear trumpet

To return to the analogy, just as that ear trumpet funnels the incoming sound waves, guiding them into an ever narrower channel, causing the energy to be concentrated and the volume within the tube to increase, so the mouth of the Severn funnels the incoming swash of the tide wave. The only difference is that the sound is a subtle succession of rising and falling pressure waves, while the bore is a single jump from low to high water.

Actually, that's not the only difference. The River Severn has Maisemore Weir to receive the wave at the far end and stop it in its tracks, while the ear trumpet has someone like me.

∿

As the Severn bore progresses upriver, its speed varies between 8 and 13mph, speeding up or slowing down according to the depth and width of the channel.

"If you see it coming and it's high, don't mess around—get back. At least it will soak you. At worst it'll pull you in." This is what the old man at the bar of the Bell Inn at Frampton-on-Severn told me the evening before I'd planned to watch the bore. "And if it's coming over the banks, don't be stupid—get out of there."

The forecast indicated that the following day's tide ought to be a particularly large one. We'd just had the new moon following the autumnal equinox, which suggested that the tidal range would be one of the greatest of the year.

The drill is quite straightforward. You stand on the riverbank to watch the bore pass but, if you're smart, you can see it at several points along the Severn's meandering course by jumping in the car and nipping from one vantage point to the next. At 8:15 a.m., I arrived at the point where the river winds around the village of Arlingham, where about fifty wave watchers had already gathered, some equipped with camping stools and thermoses. We observed the surfers, dressed head to toe in wetsuits, as they trudged out across the exposed mudflats. They carried their surfboards under their arms and climbed aboard in the narrow central channel.

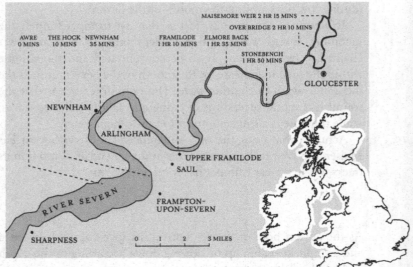

The bore progresses up the Severn at a speed of around 10mph.

I shivered as a bitter wind chilled me even through my coat, but the sight of the surfers paddling into position suddenly made me feel decidedly snug. They moved from one side to the other, anticipating the best position to catch the bore, all the while shooting glances downriver, from where the tide would arrive at any moment. The wave watchers nursed their mugs of tea, all eyes glued on the bend of the river.

You could hear the bore before it arrived. It made a bubbling, hissing sound, similar to that of a normal wave breaking, with the peculiar difference that it lacked the rhythmic ebb and flow of ocean waves, for it was *continually* breaking.

The bore front appeared as a line of white water as it rounded the bend. It didn't extend across the full width of the channel; toward the center it was a lower, smoother undulation. It only rose steeply enough to break in the shallower water at either side. Behind the bore front itself, on the higher, turbulent water level of the flood tide, followed a succession of smooth undulations. (I later learned that the members of this wave entourage go by the rather pleasing name of "whelps.")

Surf's up in about twelve and a half minutes.

Right on time, dude.

As the bore was about to reach them, the surfers turned to face upstream and, lying on their boards, began paddling as fast as they could. For some, it was all over in seconds; they were unable to pick up enough speed to catch it, or were in a part of the river where the front was either too gradual or too turbulent. But the more successful ones were away: they jumped up onto their boards and proceeded to ride the flood tide, hotly pursued by the whelps, around the bend toward Upper Framilode.

～

One of the surfers that day was Steve King. A local from the nearby village of Saul, Steve once held the world record for the longest continuous surf, a feat he performed on the Severn bore. Keen to hear from an expert about the finer points of bore surfing, I asked around and managed to meet up with him later that afternoon, outside some farm sheds that house the boat he uses for his work as a marine engineer. He surveys and maps the profiles of riverbeds and seabeds, "like the guys you see with theodolites along the road, only we do it under water." Steve showed me his vintage, sky-blue VW Camper, immaculately restored and complete with a surfboard roof rack. "I know it's a bit of a cliché," he said sheepishly.

I wasn't so sure. Along the beaches of California, yes—VW Campers clutter the parking lots. Likewise, at the surf spots of north Cornwall. But in a farmyard, halfway up a river in the heart of rural Gloucestershire his surf-mobile seemed positively eccentric. And completely in tune with the sport of surfing up a river. "It is a pretty strange experience," he admitted, "to look across to the bank and see a cow chewing grass and staring as you go by. I still find it bizarre that here, in the middle of beautiful English countryside, there's a surfable wave."

Steve first surfed the bore when he was seventeen, and has been doing so for the last twenty-six years. "I can't have missed more than a handful of tides. I've surfed almost every one." (Of course, he meant every *significant* tide. Weak bores travel up the Severn one in three days of the year, but only the highest tides offer the possibility of a ride.)

I wondered what the trick was of staying on the wave. "Well, one thing is to keep to the inside as you go around bends." One effect of the tidal current surging up the river, as well as the river current flowing the other way, is that it gouges away sedi- *Going around* ment from the outside bank of the bends, leaving the water *the bends* deeper there than on the inside. Steve surfs on the inside of the bends, since the shallower water causes the bore to bunch up and become steeper. In order to stay the distance, a surfer needs to shift from side to side of the meandering river. "You'll only know where exactly to be by studying the tides and having local knowledge because the sand shifts all the time."

How did that morning's bore compare to previous ones? "We had good rides today, with nice surfable faces. But last year was the big one. I actually broke the world distance record—I went 7.5 miles, which took a little over an hour." Alas, though he did hold the record in 1996, with a 5.6-mile ride, Steve's new distance remains unofficial. Guinness World of Records didn't recognize the claim as it was not backed up with GPS data.

The official world record of distance surfing was set by a Brazilian named Sergio Laus. On June 8, 2009, Laus rode for thirty-six minutes, over a distance of 7.3 miles, on the tidal bore that travels up the Rio Araguari in the Amazon rainforest of northern Brazil. If we count unofficial records, however, his feats are even more impressive: Laus clocked up a staggering 10.3 miles in February of the same year.

Apart from his expertise, one reason for Laus's record distances is the sheer power and scale of the Araguari bore, known locally by its indigenous Indian name *pororoca*, which pretty much translates as "bloody loud noise."

~

In the world ranking of bore heights, the 9ft one on the River Severn comes fifth.[4] Not bad—though you might have expected to see it higher up the list, given that the 45ft tidal range at its mouth is the second largest in the world. You have to remember, however, that the tidal range is just one of the three factors that determines whether a bore gets top billing. Also critical is the degree to which the estuarine coastline and river channel narrow like a funnel, and then there is the way the depth decreases toward the river mouth and along its channel.

Bore world rankings

Nor is a river's ranking permanently established. If the shape of the channel changes, the bore can disappear. The River Seine's bore used to rank second highest in the world, but it was such an extraordinary bore that it had to be sorted out. Said to have reached as high as 24ft, it caused countless boating accidents and many deaths, so in the 1960s a dredging program was implemented in which the riverbed was flattened to make the depth consistent across its span. Now *la barre* no longer bunches up steeply along the banks—in fact, the bore front has as good as vanished. When the Seine fell from grace, second place was taken by the Rio Araguari's *pororoca*, with its maximum height of 20ft.

So, if Brazil wins silver, what about the gold medal? That is held by the Qiantang River near Hangzhou in China, where the bore, known as the "Silver Dragon," or sometimes the "Black Dragon,"

Check out those lovely whelps. The *pororoca* tidal bore on the Rio Araguari,
northern Brazil, upon which Sergio Laus broke the distance surfing world record.

reaches a staggering 29ft. The tidal bores on the Brazilian and
Chinese rivers dwarf the rest of the pack. "They're of a magnitude
completely apart from all other bores worldwide," said Professor
Hubert Chanson, an expert based at the University of Queensland,
who would happily call himself a tidal bore bore.

But even as the medals are being handed out, there is contro-
versy: although the *pororoca* is lower than the Silver Dragon, and
travels more slowly up the Araguari (at 15mph, compared with
the Qiantang's 25mph), it is, according to the experts, the most
powerful bore on the planet. "The energy dissipated by the *pororoca*
is twice that of the Qiantang," claims Chanson. "It is a much wider
process and propagates much further—lasting in excess of twelve
hours, compared with the 2–4-hour duration of the Qiantang
bore."

Whether or not it is the most powerful, the Silver Dragon has
a deserved reputation as the most dangerous bore, having claimed
the lives of around one hundred wave watchers in the last two

decades alone. The volume of the flood tide on the Qiantang may not be as great as that on the Araguari, but the steep funneling of the 60-mile mouth of the Hangzhou Bay down to a width of just under 2 miles forms a dirty, angry, turbulent wall of water of incredible ferocity.

One can only imagine the death toll over the millennia, for this form of wave watching has a long history. Visitors have been gathering to watch the Silver Dragon since at least AD 1056, when *Bored to death* the world's earliest known tide timetable was compiled. It was produced especially for bore watchers, and listed the optimum dates and times to follow the tide's progress up the river.[5] But it was probably an attraction long before then, as there was clearly already enough demand for comfortable seating around that time to warrant erecting a bore-watching pavilion at the riverside town of Yan'guan.[6] Furthermore, the philosopher Zhuangzi had described the dramatic tide in the fourth century BC:

> *The waters in the river will roll on, raising waves as high as mountains and towers, creating a thunderous roar and gathering up a force that threatens to engulf the sun and the sky.*[7]

Threats to engulf sun and sky might have proved somewhat empty, but the bore certainly threatens to submerge anyone standing too close to the bank. The worst tragedy of recent time occurred on October 3, 1993, when eighty-six people were swept away. As a result, the local government established designated tide-viewing areas for the China International Qiantang River Tide-Watching Festival, which takes place during September, when the bore is highest.

Yet accidents continue to happen. It is not that people are caught unawares—in the eleventh century the poet Su Shi compared it to "ten thousand men's roar / As if downstream came conquerors' warships and glaives."[8] But visitors don't appreciate the violence with which the bore can rear up as it sweeps by. If the water is shallow near the bank where they are standing, the bore can suddenly bunch up and steepen without warning, crashing over the river wall and inundating the walkway. The mood of the Silver Dragon can change in an instant.

" 'Tis best to watch the tide in the moonlight," continued Su Shi,[9] but you can forget about that nowadays. Nocturnal wave watching has been banned following yet another, more recent, incident, on August 2, 2007, when thirty-four spectators were swept off into the darkness. Clearly, when the Silver Dragon is hungry for wave watchers, it will claim them, day or night.

Since the entire globe experiences a gravitational pull from the same sun and moon, why do tidal ranges differ so much? Why does Tahiti, in the middle of the Pacific Ocean, experience a tidal range of less than 12in, while Brisbane on the Australian coast, 3,700 miles due east, sees 5.5ft? Why does the Severn estuary have a maximum tidal range of 45ft, when Newquay, just 100 miles along the coast, has half that? The answers are never straightforward, and they are a combination of global and local factors.

Tide waves don't just travel from one side of the oceans to the other, east to west and back, as you might expect, given that the earth rotates in an easterly direction. In fact, the high water of the tide waves rotates around the edges of the oceans. To understand what this means, try a simple experiment. Take a frying pan filled with an inch of water and, as it sits on the kitchen surface, slide it gently from side to side so that a wave builds *Ocean basins and frying pans* up, rising on one side, then the other. This is *not* how the tide waves move in the ocean. To make your frying-pan wave travel in a more analogous way, introduce a slight rotation: slide the pan in gentle circles, so that the wave travels *around* the rim. This is more like it. The tide waves tend to rotate around the ocean basins so that the high tide, the crest of the tide wave, reaches one port after another.

Some regions of water, known by oceanographers as "amphidromes," are at the center of a system of rotating tide waves. Here, as in the middle of the frying pan, the water level barely rises and falls. Where an island is located near an amphidrome, as is Tahiti in the Pacific, for instance, it will experience negligible tides.

But naturally, it is not as simple as that. The ocean floors are not flat, smooth and coated in Teflon. In fact, seafloor and coastline

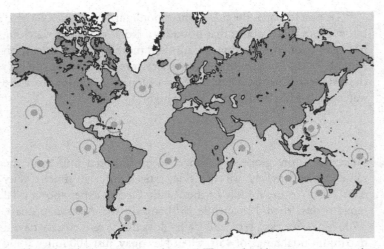

Think the tides just come in and go out? Wrong: the tide waves rotate around amphidrome points, the main ones of which are shown here.

irregularities mean that the rotation of the tide waves is a lot more complex. The water develops rotations around different regions within the basin.

But what sets the tide waves rotating like this? It is the spin of the planet. There is a tendency, known as the "Coriolis effect," for any large-scale movements of water over the surface of our spinning globe to be deflected sideways.* It is the same effect that explains storm systems rotating in the familiar spiral patterns we see on satellite images. The Coriolis effect accounts for water that is moving in the Northern Hemisphere to be deflected toward the right and in the Southern Hemisphere toward the left. This is why the tide waves generally rotate clockwise in parts of the ocean north of the equator and counterclockwise in those south of it.

So, the rotating nature of tide waves explains some of the global variation in tidal ranges—according to the coast's proximity to an ocean's amphidrome—but why are there local differences in tidal ranges? For local reasons, unlike your frying pan, the coastline is

* It does not actually change direction, but *appears* to from the point of view of anyone, like us, also on the rotating globe.

never flat and smooth. As we've seen, the funneling effect of bays and estuaries has the effect of bunching up the incoming tide, and so increasing the tidal range, compared with the neighboring regions of the coastline.

Another factor that will increase tidal ranges along one part of a coastline compared with another is the effect of "tidal resonance." This is when the reflections of one tide wave add to the crest of the succeeding tide wave, causing the range to be greater than it otherwise would be. A tide wave reflects off the coast, just *Our old friend* as a ripple on your bath water reflects off the side of the *resonance* tub. When a coastline is surrounded by a continental shelf, as is the case around most ocean basins, a system of tidal resonance can develop as a result of the way this reflection interacts with the subsequent low tide. If the distances happen to be just right, so that the tides and reflections meet at the shelf edge, the resulting tidal ranges at the shore can be the largest in the world, such as in the Bay of Fundy, in Canada, and in the Bristol Channel, at the mouth of the River Severn.

From the head of the Bristol Channel there is a distance of about 400 miles into the Atlantic to the point where the continental shelf of Europe ends and the seabed slopes down into the murky deep. It just so happens that the sudden change in depth at the edge of what is known as the Celtic Sea Shelf is a distance of about a quarter of the wavelength of the principal tide in that part of the Atlantic. This means that a standing wave, or seiche, develops with a node (where rise and fall in water level is minimal) at the shelf edge, and an antinode (with large rise and fall) at the head of the Bristol Channel (see next page).

However minor the reflections off the coast, they still build up a resonance in the movement of water since, as with pushing someone on a swing in time with their natural arc, the reflections here always combine in sync with the subsequent low tide to form a node at the shelf edge. The resonance causes the rise and fall of the water to build up at the coast, amplifying the tidal range compared to how it would be without a continental shelf extending that particular distance from the shore. The same principle contributes to the even larger tidal range in the Bay of Fundy.

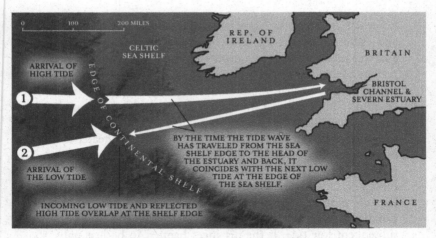

ABOVE: The high tide reflects off the head of the Bristol Channel and meets the next low tide at the edge of the Celtic Sea Shelf. This helps develop a tidal resonance (BELOW) with a node at the shelf edge and an antinode near the river mouth.

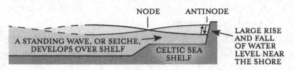

Some seas, by contrast, experience nothing but the feeblest of tides. This is simply because they are not large enough, for the smaller the area of an enclosed sea, the less pronounced will be its tides. Remember: the tide-inducing gravitational forces depend on the *difference* in pull between one side of a body of water and another. Only when the sea or ocean is large is this a pronounced difference. This is why the tides in the Baltic and Mediterranean are so minor as to be almost unnoticeable.

Alexander the Great found himself in a spot of bother with the tides during his invasion of the Indian subcontinent in the fourth century BC. Having only ever experienced the minimal tides of the Med, that colossus of ancient history was completely unprepared for what happened when he moored a fleet of light ships on the River Indus, which flows the length of what is now Pakistan. According to the Greek historian Arrian, the receding tide left the vessels high and dry on the mudflats, to the bafflement of

Alexander's men. The real shock, however, came later in the day, when the boats were refloated with the surging flood tide, which traveled up the river as a bore:

> *The ships which it caught settled in the mud were raised aloft, without any damage and floated again without receiving any injury; but those that had been left on the drier land and had not a firm settlement, when an immense compact wave advanced, either fell foul of each other or were dashed against the land and thus shattered to pieces.*[10]

The same lack of tidal experience explains why the Bible contains absolutely no reference to tides. Palestine's coast is Mediterranean and the Israelites were not a seafaring people.

While geographical factors make tide prediction difficult, the task is further complicated by how the tidal range at any place varies over time. The main reason is the changing positions of the sun and moon. The largest tidal ranges, spring tides, happen at *Spring and* full or new moons. This is when the sun, the moon, and *neap tides* the earth are in line, so that the gravitational pulls of the sun and moon upon the earth are lined up. Neap tides, on the other hand, when the tidal range is at its least, occur when the positions of the sun and moon mean that their gravitational pulls are not in line, but at right angles to each other. When they are positioned like

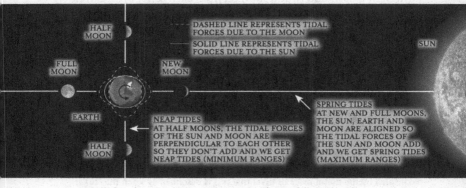

Positions of earth-sun-moon at spring/neap tides on full, half, and new moons.

this, we see the sunlight falling on the side of the moon, so it appears as a half-moon. The moon's tidal influence may be greater than that of the sun, but the most important thing is how the two combine, as determined by their relative positions.

Moreover, the tidal range also depends on the weather. The low air pressure of a storm system means the air is not pushing down on the water quite as much as elsewhere, so the water level is raised slightly. When this effect combines with storm winds blowing ashore, the resulting rise in water level is known as a "storm surge." Flooding is a serious risk in coastal regions on those occasions when a large storm surge coincides with high tide. The storm surge from Hurricane Katrina, which hit the southern coast of the United States in August 2005, was responsible for the devastating flooding in and around New Orleans. The surge raised water levels as high as 28ft, and in some areas combined with the high tide to reach over 30ft.

And the weather is not even the end of the causes for tide variability. Also to be factored in are the changing distances between the earth and the moon through the lunar month, and between the earth and the sun through the solar year. The gravitational forces wax and wane, and are at their greatest when the earth happens to be simultaneously closest to both the sun and moon.*

No wonder it is practically impossible, even with modern computers and water-level gauges, to accurately predict the tides from one week to the next, from one coastline to the next. The different factors add to and subtract from each other in an extremely complex equation of forces upon the oceans. But the end result is simple: the ceaseless rise and fall of water.

～

Clearly there is a huge amount of energy within the movements of colossal bodies of water by the tides. Given our need to move away from fossil fuels, it would seem to make sense to capture some of

---

* Those dejected pub-quiz aficionados I imagined back at the bar might be consoled by the useful piece of trivia that this is described as when the "lunar perigee" coincides with the "perihelion of the earth."

that energy for ourselves. The technology exists to extract power from both wind-generated ocean waves and gravity-generated tide waves, and the UK with its extensive coastline and high tidal ranges, is perfectly situated to exploit both forms. A government report in 2006 estimated that 15–20 percent *Tide power, the future?* of the British nation's current energy needs could be provided by "marine renewable energy."[11] If that sounds far-fetched, in the late 1990s the government's Marine Foresight Panel had worked out that "if less than 0.1 percent of the renewable energy within the oceans could be converted into electricity it would satisfy the present world demand for energy more than five times over."[12]

Yet there's nothing newfangled about tidal power. In 1999, the ruins of a tidal mill were uncovered near the medieval monastery of Nendrum, on an island in Strangford Lough, Northern Ireland. The lough is a tidal lake with a range of around 12ft. During flood tide, the monks would have filled the millpond, letting the water in through a sluice. Once the tide had receded, the pond water was diverted down a stone channel onto a horizontal paddle wheel that turned the millstone to grind flour in the building above. Dendrochronology of an oak beam found below the millstone during the excavation revealed it to have been from a tree felled in AD 787.[13] If, as archaeologists believe, the mill was built around this time, it is the earliest excavated tide mill in the world.

While tide mills were never as common as wind or river mills, there is evidence of at least 220 having operated at one time or another in England and Wales.[14] Only seven still stand, two of which are in working order. On the east coast of the United States, around three hundred tide mills are thought to have existed, and on France's Atlantic coast perhaps one hundred.[15] Two such early "tide power plants" were still in operation at the mouth of the River Rance in Brittany when work began there to construct the world's first commercial tidal power electricity generating facility. Completed in 1967, it has been generating a net power of 540 gigawatt hours (GWh) every year—enough to supply 250,000 households.

The Rance tidal power plant consists of a 2,500ft barrage across the estuary. During the flood tide, the seawater turns twenty-four turbines as it flows inland. On the turn of the tide, the sluice gates are closed and the water held back. It is then released to power the turbines once again, on its return to the sea. The plant can also be used as an energy storage facility: when there is a surplus of energy in the grid, this can be used to pump water up into the river basin for conversion back into electricity when the demand is higher. With over forty years of continuous, trouble-free, carbon-neutral service, the scheme is about as green as it gets.

~

So, given that the Severn Estuary experiences such a dramatic tidal range, why haven't we bunged a barrage across it?

In 1920, the civil engineering department of the Ministry of Transport published a plan to do just that, proposing to do so because of "the high price of coal and . . . the labor situation in respect of coal-getting."[16] The idea was met with skepticism and hostility by much of the press, leading further investigations to be postponed. But in 1927 the government took the idea a step further, appointing a committee to find a suitable site along the estuary. Fears for the jobs of the South Wales coal miners led to the scheme's deferral. When it was again shelved in 1945, the numbers didn't stack up: electricity generated by a barrage, though cheaper than existing coal stations, would still have cost more than from the new, state-of-the-art coal power plants. Nor did a barrage seem feasible in 1981, when yet another government committee found that, although it was cheaper than coal, tidal power would be more expensive than nuclear power.

Since the new millennium, the tide has turned once again. This time the arguments both for and against are environmentally focused. The UK has made a commitment to the EU to increase the proportion of electricity generated renewably from the current figure of 5 to 15 percent by 2015. A recent report estimates the power that could be generated by a barrage spanning the 9 miles between Cardiff and Weston-super-Mare as around 17 terawatt hours (TWh), equivalent to that of three nuclear power stations.[17] Such a scheme, containing 216 water turbines, would be enormously ambitious and expensive to construct, costing in the region of $22.5 billion.

You'd think this would be a no-brainer from a green point of view, but the thought of such a scheme has horrified many conservation and environmental groups. The Royal Society for the Protection of Birds is against it, for instance, as it would flood the tidal mudflats used as a feeding ground by 69,000 migratory *Green objections* and wintering birds each winter.[18] Friends of the Earth also object to its monolithic nature, arguing that the barrage would not only draw investment away from other renewable energy technologies, but would produce huge surges in supply, which would be difficult and costly to feed into the national grid.[19]

Others have argued that the changes to the flow of tidal water in and out of the sediment-laden waters of the Severn would cause considerable silting of the river channel, increasing the risk of floods during heavy rain in Gloucestershire, an area already prone to them. They point out that the lack of silting upriver of the Rance tidal power plant is thanks to the river water there carrying relatively little sediment. By contrast, the causeway constructed in 1968 across the sediment-heavy Petitcodiac River in New Brunswick, Canada, and which opens onto the Bay of Fundy, has caused large areas of the river to silt up, with disastrous results for fish and other wildlife. The scheme, which merely enabled a roadway to cross the river, was described by the *Montreal Gazette* as "among the stupidest" of the "atrocities mankind has inflicted on his environment."

The global economic downturn may again have sealed the fate of the Severn Barrage. In October 2009, the *Times* reported that ministers were questioning the project's affordability, and quoted a Westminster insider: "They are moving toward a political fudge. They will say they are delaying it, but in reality the lifeline on offer will not be worth very much."

If history is anything to go by, we haven't heard the last of it. Next time, should the barrage actually end up being built, there will be one other catastrophic outcome: it will put an end to the bore. Never mind all the other environmental stuff: if the barrage gets the green light, there will be bore surfers and wave watchers to contend with. And the latter will come armed with thermoses of tea.

It is just possible that the tides have been instrumental in the birth of life on the planet. Most experts agree that around the time Earth was forming—some 4.5 billion years ago—our proto-planet was blindsided by another, called Theia, which was about the size of Mars. This impact, sometimes referred to, in all seriousness, as the "Big Whack," threw into space a blanket of molten matter, consisting of some of the earth's mantle as well as the remains of Theia. In just a year, this had coalesced into a molten ball, the moon.

Fast forward half a billion years and our planet had developed oceans and had cooled enough to form a solid crust. But it was devoid of life.

In its youth, the earth spun much faster than it does today, meaning the days were much shorter. There is no consensus about how short, but some have estimated just fourteen hours, which would have *Shorter days,* meant high tides every seven hours or so.[20] Around this *swifter tides* time, the moon was much closer to earth, and so its gravitational pull on the oceans much greater. Therefore, the tides before life began were far more powerful than they are today. Oceans would wash over the early continents at speeds of up to 300mph, scouring the ground as they did so and washing minerals into the water, which would be essential for feeding any future life.

But some scientists have argued that the early tides might have played an active role in creating life in the first place. "A lot of origin-of-life reactions involve getting rid of water," Kevin Zahnle, a planetary scientist at the NASA Ames Research Center in California, has explained. "So you look for means to concentrate your solutions. One way to do that is to throw water up on a hot rock, then have the waters recede and evaporate."[21] This, of course, is a role that tides would have performed with ease.

Thinking along the same lines, the molecular biologist Professor Richard Lathe has argued that the regular ebb and flow of water over enormous tracts of barren land might have served as a driving mechanism to multiply early versions of DNA and RNA, the genetic code-carrying molecules. The tides might have served to multiply these molecules in a way that is similar to that used by modern forensic scientists to replicate DNA.

When forensics are asked to make a DNA profile from a tiny sample gathered at a crime scene—perhaps something as insubstantial as a hair follicle—they need to reproduce the DNA to obtain a large enough sample to test. This they do using a "thermal cycler," which repeatedly heats and cools the molecules, multiplying the DNA by continually breaking and reforming their helical structures. (If you really must know, this is called a "polymerase chain reaction.") The same mechanism of reproduction could, Lathe suggests, result from regular cycles of drying and wetting with salt water. In short, the early tides might have orchestrated the reproduction of the building blocks of life.[22,23]

So just remember the debt we may possibly owe the tide the next time it washes away your award-winning sandcastle.

Over the billennia, the tides have caused the moon gradually to move away from us. Were there no tides, the force that the earth exerts on the moon would pull it directly toward its center. But the presence of the tides causes a fractional displacement of the earth's gravitational pull. Due to the earth's spin, the mass of water in the tidal bulges is always slightly out of alignment with the moon, which has the effect of very gradually slinging it outward into an ever-larger orbit. As a result of the tides, the moon is currently receding at a rate of 1.5in a year.

Meanwhile, the energy dissipated by the friction of huge quantities of water sloshing around in the ocean basins has also had the effect of slowing the spin of the planet. In this way the fourteen-hour days of 4 billion years ago have gradually stretched to the twenty-four-hour ones we know and love.

So if you ever feel that life is moving too fast, and you never get the stuff done that you hoped to, fear not, for the tides are working in your favor, continuing to slow the earth's spin, forever buying you a little more time before *Good news for the habitually late* sundown. Fifty years from now, thanks to these waves, you'll have gained an extra 0.001 seconds in every day.[24]

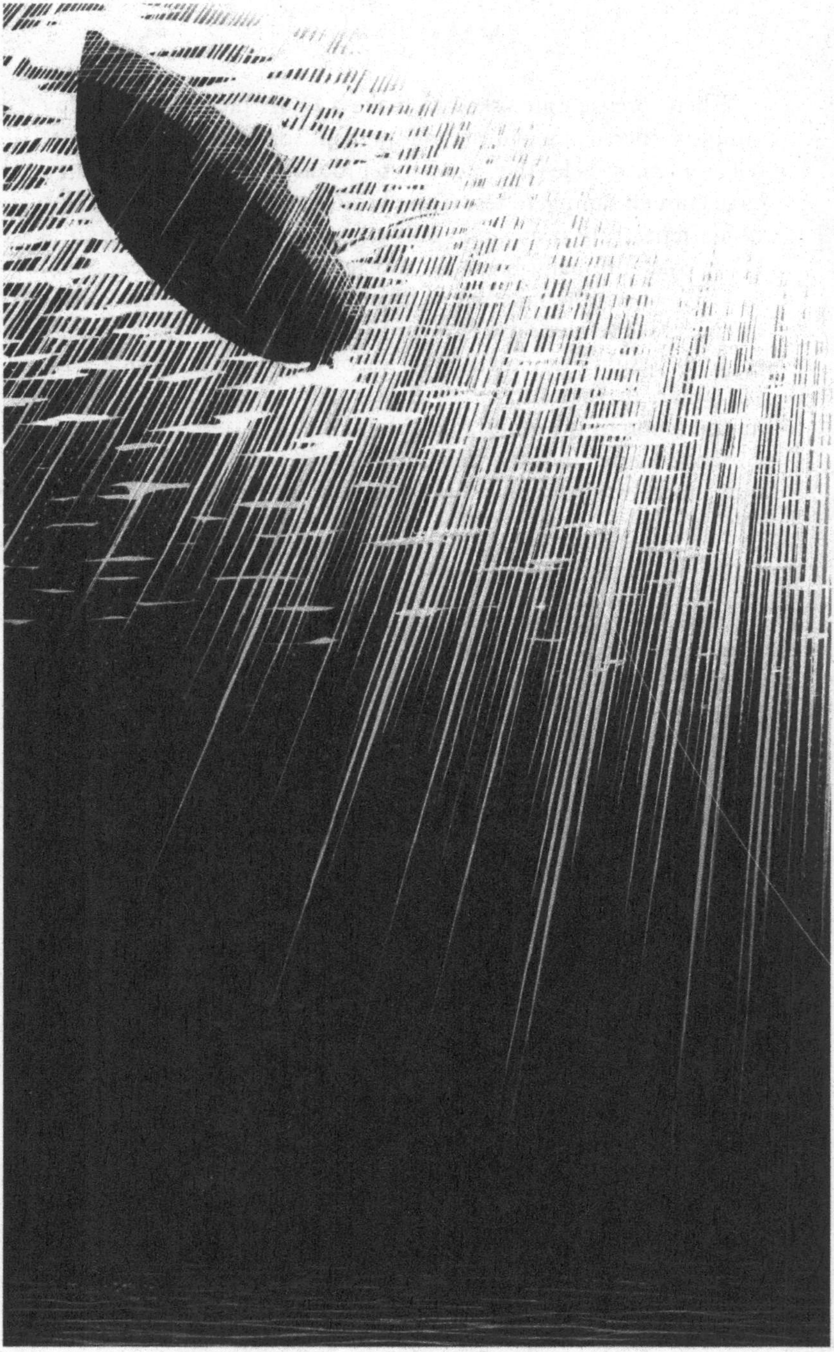

The Eighth Wave

WHICH BRINGS COLOR TO OUR WORLD

On a walk in the woods one October afternoon I came across a peacock butterfly. It was easy to recognize on account of the four distinctive "eyes" on its wings, which resemble those on the feathers of its flamboyant avian namesake. When the butterfly settled on the bark of a beech tree in a patch of dappled autumn sunlight, it opened its wings to gather what warmth it could. It must have been feeling the chill, for, as I stepped up to take a look at its markings, it didn't budge.

The blue patches within the eyes seemed to jump out from the burnt orange and dusky yellow on the rest of its wings. These patches had an iridescent quality, by which I mean that the color shimmered and changed slightly with the angle, like one of those silk scarves on which the hue shifts subtly around its folds. Compared with the flat colors surrounding it, the blue looked almost three-dimensional.

In 1634, Sir Theodore de Mayerne, physician to Charles I, wrote that the peacock butterfly's eye formations "shine curiously like stars, and do cast about them sparks of the color of the Rainbow." Although we may find them beautiful, the patches have evolved not to allure but to alarm. The blue creates the optical illusion with which the butterfly defends itself from predators. If the bark-like wing undersides don't camouflage it from a passing woodmouse, the butterfly will flash the upper sides in an attempt to startle the furry predator. To the rodent, the eyes on the wings resemble the face of a fierce and hungry owl.

Butterfly iridescence

Like those of all butterflies, the peacock's wings are covered in tiny scales, each one around 0.2mm long and 0.075mm wide. These are what give the wings their color. The orange and yellow scales, along with the black and white, are colored by pigments. These are chemicals that efficiently reflect certain hues of light, while absorbing the others. But the iridescent blue of the "eyes" are not caused by pigments; although there *is* pigment, called melanin, within these scales, it is just a dull brown color. The secret lies in a transparent material, known as chitin, which covers the scales.

How can an iridescent blue shine out from a scale containing a dull brown pigment covered in a *see-through* surface? The blue, unlike the other colors, is known as a "structural color." This means that it is produced by the physical structure of the transparent surface, which consists of extremely fine layers separated by minuscule gaps. Each of the layers reflects some of the light falling onto them. The glorious blue is the result of the way light reflecting off the different layers combines, and this is thanks to a phenomenon known as "interference."

~C~

Interference is what happens when waves of the same type collide, and it can happen to all waves, not just light. Actually, "collide" is not the right word; rather, they pass through each other and add together as they overlap. I saw a fine example of wave interference in my daughter's paddling pool. The waves were produced not by some flashy butterfly, but by a couple of very unhappy moths.

How they came to be there was unclear, but overnight they had become trapped on the surface of the water. Their mottled, fawn-colored wings were stuck to it, pinned down by the surface tension. Flapping their wings, they now produced a steady succession of ripples across the surface. The effect of the overlapping ripples was intriguing, and I found myself torn between the desire to set the moths free and to observe them.

I could tell from the matching wavelengths of the circular ripples that they flapped their wings with similar frequencies. At one point, the two moths drifted particularly close to each other and the little waves they produced combined to form a beautiful interference effect. As the ripples spread outward from each stricken insect, they created a pattern of lines fanning out like spokes from the gap between them, along which the two sets of waves either added to make extra large ripples or subtracted to cancel each other out. The water looked rather like this:

Moth interference

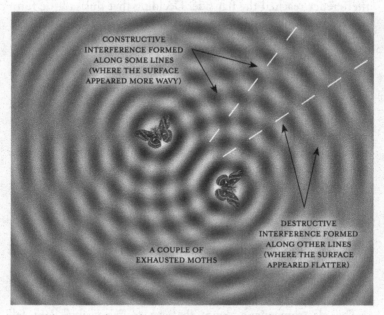

Beautiful wave interference from the desperate flapping of moths imprisoned on the water of my daughter's paddling pool.

The lines of relatively calm water were where the crests from one moth coincided with the troughs from the other. Being out of phase like this, the waves canceled each other out, which is to say they interfered "destructively." In between were lines where the water was much wavier. This was where the ripples arrived in phase with each other, the crests from one moth always meeting the crests from the other, and the troughs meeting the troughs, so, as they added together, they interfered "constructively." Only when two sources of waves are "coherent" with each other, meaning that they share the same frequency and heights and don't change phase relative to each other, do they produce a fixed pattern of interference like this, so I suppose I wouldn't have seen the same effect from moths that flapped their wings at different speeds.

Along the nearest edge of the rectangular pool I had just about been able to distinguish alternating patches of bumpier and less bumpy water where these lines of constructive and destructive interference reached the flat side. Deciding that this sadistic form of wave watching had got out of hand, I'd scooped the moths out and let them dry on the decking to fly another day (or night).

When you are flying, it is sometimes possible to see wave interference patterns on a much larger scale. From the window of an aircraft flying over the ocean, you can sometimes see the regular lines of a swell that is coming from one direction overlap with those of a swell from another direction. They merge to form a cross-hatched pattern of combined and canceled crests and troughs. In fact, all waves of the same type interfere like this when they pass through each other—unless, that is, they are shock waves, when the normal rules break down.

Interference is such a fundamental behavior of waves that it should come as no surprise to learn that light waves exhibit it too. But it takes a little explanation to show how it can explain the iridescent color on butterfly wings.

The electromagnetic waves that we see as visible light fall within a range of wavelengths—from somewhere around 400–450 nanometers (nm),* which appears indigo and blue to us, up to around 700–750nm, which appears red. The numbers are somewhat vague as it is hard to say exactly where visibility ends, *Visible and* depending as it does on the conditions and who is doing the *invisible waves* looking. Just as pit vipers of Central America are adapted to see the longer, infrared wavelengths given off as body warmth by their prey, so bees, for instance, can see the shorter, ultraviolet wavelengths reflected by certain flowers. Both are invisible to us, however.

The fact that the wavelength of light determines its apparent color is crucial to understanding the iridescence of a peacock butterfly's wing. The way structural, non-pigmental colors like this are formed by interference depends on the fact that light waves with a distance of 400nm between their peaks look blue.

In order to explain how structural colors work, it might be easier to refer to another kind of butterfly, one whose iridescence is so dazzling that it has been the subject of a great deal of scientific study. *Morpho* butterflies are found in the jungle canopies of Latin America. Some species within this order have iridescent blue across their entire wingspan, which can reach an enormous 8in. So intense are the flashes of blue as they flap their wings (flutter seems too delicate a term for such large specimens) that they can be spotted from a quarter of a mile away, and be seen glinting above the treetops by low-flying aircraft.

The structures on the surface of *Morpho* wing scales, like those responsible for all butterfly iridescence, produce iridescent colors as a result of sunlight reflecting off incredibly fine, repeating layers of transparent chitin. These are far too fine to be seen through a normal, optical microscope. In order to reveal the secret of the *Morpho*'s mesmerizing blue, you have to photograph the scales through an electron microscope. Observed in this way, the separation between the cuticle layers turns out to be a staggeringly uniform and minuscule distance: some 200nm, which, as it happens, is about half the wavelength of the blue light.

Take the *Morpho rhetenor* butterfly. Ridges of the transparent cuticle run along the length of its tiny, iridescent wing scales, each

---

* A nanometer is, if you remember, a thousandth of a thousandth of a millimeter.

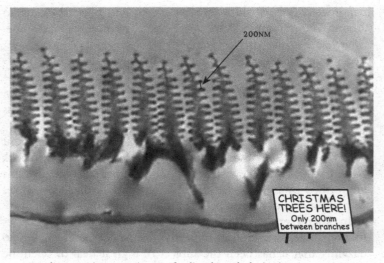

An electron-microscope image of a slice through the intricate structure on the surface of the *Morpho rhetenor* butterfly wing scales reveals it to look like the outside of a garden center in the run-up to Christmas.

of which is smaller than a full stop. Along the sides of each of these run even smaller ridges. When you use an electron microscope to photograph a cross section of the surface of this butterfly's wing scale, the ridges resemble tiny Christmas trees.[1]

But even if the Christmas trees were up at human scale, you would never be able to fit them in your living room because each "tree" is just a slice through a ridge, extending along the length of the scale, rather like transparent Play-Doh that has been squidged through a Christmas-tree-shaped extruder. The "branches" are mini-ridges running along the sides of the main ones, and are the all-important layers, spaced with such astounding precision—each one exactly 200nm from the next.

The light waves reflecting off the upper and lower surfaces of these layers of cuticle interfere with each other as they overlap. Sunlight bouncing off the top of a layer interferes with light that has passed through the transparent material and bounced off the bottom of the layer. How the two reflections interfere depends on the difference in the distance that each light wave has traveled, the change in speed when passing through the cuticle, and the wavelength of the light. The comparability of these measurements determines whether the two overlapping light waves are in phase with each other, so that their crests and their troughs coincide and add up, or out of phase, so that the crests coincide with the troughs and they cancel each other out. And, depending on how in phase or out of phase the waves are, they will either combine constructively to appear brighter or destructively to appear dimmer.

The extremely precise thickness and spacing of the Christmas tree branches ensures that, of the whole spectrum of wavelengths found in sunlight, only wavelengths close to the 400nm-long blue light interferes constructively and shines brightly. Because the reflections of the blue light are all in phase with each other, they build in intensity as they bounce off the Christmas tree branch layers, while those of the other colors are out of phase and interfere destructively, canceling each other out and appearing dim. Successive layers amplify this wavelength-specific reflection and the brown melanin pigment at the base of the Christmas trees

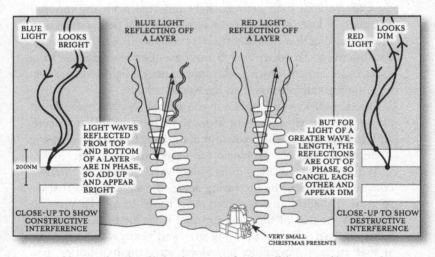

For blue wavelengths of light, the waves reflecting off the top and bottom of a layer are in phase, so they appear bright. Longer wavelengths, such as red light, are not in phase, so they appear dim.

absorbs any remaining wavelengths of light that haven't reflected off them, thereby preventing these from washing out the purity of the blue. In this way, the phenomenon of interference acts like a magical wave selector, teasing out from the tangle of different wavelengths (which appear to us, in combination, the color of sunlight) just the narrow band of blue that shines with such an intoxicating luster.

To see a *Morpho* butterfly in a Victorian display cabinet is rather to miss the point. The butterfly's beauty lies in the way the colors change as it opens and closes its wings. If you don't have a rainforest to hand, visit the butterfly house of a zoo and you will see that the electric blue color, besides being astonishingly vivid, subtly shifts to an indigo blue as the butterfly angles its wings. When you are no longer looking at the wings face-on, the hue shifts. This irides-cent effect, the reason why the color feels infinitely deeper than a normal pigment, is one that comes alive with the movements of the insect.

Again, it is the interference that causes these subtle shifts in hue. When light falls onto the cuticle layers at an oblique angle, the difference in path between light reflected from the top and bottom of a layer is not so great as when it falls perpendicular to them, meaning that a slightly shorter wavelength of light interferes constructively. So when viewed from an oblique angle, the wing looks more indigo, which is the appearance of light with a wave-length marginally shorter than the electric blue.

Another effect you'll notice if you see the *Morpho* butterfly up close is how the glorious color seems to flash on and off as it opens and closes its wings. When you look at the wing from a very glancing angle, almost side-on, the blue color disappears altogether.

While it is the shorter, indigo-blue wavelengths that interfere constructively with the increasing angle, the wavelengths that do so at *really* shallow angles are too short for us to see. They are no longer in the part of the spectrum that is visible to us. The color seems to flash off when the wing is viewed nearly side-on because at that angle the invisible, ultraviolet wavelengths (shorter than 400nm) are then the ones that interfere constructively and "shine." With the wing movements in flight, the blue color is constantly

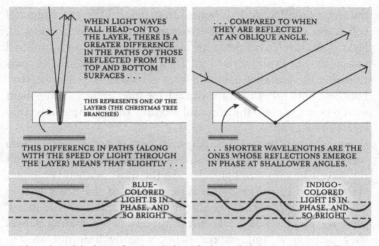

WHEN LIGHT WAVES FALL HEAD-ON TO THE LAYER, THERE IS A GREATER DIFFERENCE IN THE PATHS OF THOSE REFLECTED FROM THE TOP AND BOTTOM SURFACES . . .

. . . COMPARED TO WHEN THEY ARE REFLECTED AT AN OBLIQUE ANGLE.

THIS REPRESENTS ONE OF THE LAYERS (THE CHRISTMAS TREE BRANCHES)

THIS DIFFERENCE IN PATHS (ALONG WITH THE SPEED OF LIGHT THROUGH THE LAYER) MEANS THAT SLIGHTLY . . .

. . . SHORTER WAVELENGTHS ARE THE ONES WHOSE REFLECTIONS EMERGE IN PHASE AT SHALLOWER ANGLES.

BLUE-COLORED LIGHT IS IN PHASE, AND SO BRIGHT

INDIGO-COLORED LIGHT IS IN PHASE, AND SO BRIGHT

The color of the butterfly wing shifts with the angle it is viewed at as different wavelengths interfere constructively. I tried to make this diagram simple, but I admit that there is a faint possibility that I failed.

turning on and off. There is an evolutionary advantage for *Morpho* butterflies to produce a flashing blue light like this: it is as alarming to predatory birds as it is to a speeding motorist when seen through the rear-view mirror on the motorway.

∼c∼

Iridescent, structural colors don't appear only on butterflies. The wing cases of beetles, for instance, exhibit a fantastic palette of shifting metallic hues. In the case of the remarkable Japanese jewel beetle (*Chrysochroa fulgidissima*), these appear on its underside as well as on its wing cases. As you observe it from different angles, it changes from yellowish-green to deep blue on top, and from green to rich reddish-brown below. You can see iridescence on far less exotic species, too. The tiny ¼in-long mint leaf beetle (*Chrysolina menthastri*) is a rich green tinged with copper. It may decimate your garden mint, but at least it does so with panache.

Iridescent fashions for beetles

The colors of iridescent beetles are not caused by structures that look like Christmas trees in cross section, but are based on the same principle of light-wave interference. Here the structural colors are also produced by transparent chitin, which is spread, layer upon layer, over the wing cases, each surface just a hundred nanometers or so above the next. Their iridescent jades and coppers resemble the metallic colors of thin oil slicks on water or soap bubbles, both of which are caused by interference of the light waves bouncing off the top and bottom of the incredibly thin layers of oil or soap.

And then there are birds. Any feather with an iridescent sheen will, to some extent, be the result of structural colors that exploit the wave-like nature of light. Perhaps the most dramatic examples are the vivid blues, greens, reds, and golds of the many species called birds of paradise. The intensity of the colors on the males' plumage plays a major role in the success of the mating dances they perform to impress the ladies. Equally beautiful are the shifting tones on the necks of certain hummingbirds. Greens and blues prevail, but some species employ iridescent purples, yellows and copper-oranges. The distinctive turquoise flash of a kingfisher comes courtesy of structural colors, as does the blue/green around the neck of a common male pheasant. And let's not forget the eyes on the iconic peacock feathers.

The precise structures that create the light-wave interference to cause these glorious colors vary between species of bird. A peacock feather consists of a central spine, with many barbs running off it. Each of these has countless smaller branches, called barbules. When viewed under an electron microscope, these barbules are found to contain "photonic crystals," which consist of nanoscale 3D lattices of melanin granules spaced at intervals comparable with the wavelength of light.[2,3]

For all these creatures, and the many others like them, to have developed such intricate surface structures, the iridescent colors they produce must have brought considerable evolutionary advantages, such as communication—whether between friend or foe. But one beautiful example of iridescence—the pearlescent colors produced by the layered "nacre" that lines the inside of an oyster shell—seems to be nothing more than the side effect of creating a safe home. Mother-of-pearl consists of nanoscale plates of calcium carbonate glued together in countless layers to make a smooth, shatterproof surface. Hidden from view to anyone but the mollusk itself, it can play no role in reproduction or communication. That said, it would be nice to think that oysters are happier and healthier for having attractive wallpaper.

Wherever humans are around, however, iridescent colors have been unhelpful for the creatures concerned. The wings of *Morpho* butterflies make rather fine decorations on ceremonial masks—or so the Amazonian tribes that trapped them have found—and human beings have tended to appreciate mother-of-pearl even more when inlaid on a wooden cabinet than inside a living mollusk's shell. In the mid-nineteenth century, aristocratic European ladies loved to show off by wearing ballgowns appliquéd with the wing cases of jewel beetles. Such beguiling structural colors may have served the creatures well in wooing mates and repelling predators, but their appeal to human beings must rather have taken the shine off things.

Iridescent fashions for humans

⌇

"We all know what light is; but it is not easy to *tell* what it is," observed Samuel Johnson.[4] He's right: the fact that light *enables* our vision makes it a formidable challenge to resolve, with any clarity, the nature of light itself.

I must make a confession. I've been describing light as a form of wave but it is not straightforwardly so. Robert Hooke, the seventeenth-century English physicist—or "natural philosopher," to use the designation of the time—proposed a wave theory of light in 1665. The theory was strengthened some twenty-five years later, when his Dutch contemporary Christiaan Huygens published mathematical proof that much of the behavior of light could be explained in terms of waves.

Light is made of waves

The one problem with the theory was accounting for what the light waves were passing *through*. Ocean waves travel through water, sound waves through air (or any other material), but what was "waving" as light passed through? Since all other waves require a medium, and light can pass through a vacuum, the wave theorists needed to propose some sort of luminiferous, or light-bearing, ether—but no one could say what this ether might consist of.

Sir Isaac Newton suggested a rather different concept of light in his seminal book *Opticks*, first published in 1704. He suggested *Light is made* that it might consist not of waves, but of tiny particles, or *of particles* "corpuscles." And the idea rather stuck, for *Opticks* became the defining work on the behavior of light for the entire eighteenth century. (For such an important book, you'd think that he might have spell-checked the title.)

Newton demonstrated with ingenious experiments and skillful deduction how and why light behaves as it does when diffracted by glass, famously showing that sunlight can be divided by a prism into its constituent rainbow spectrum of colors. The corpuscle theory was by no means central to the book. In fact, it was only mentioned in a "Query" at the end of the revised, 1717, edition: "Are not the rays of light very small bodies emitted from shining substances?"[5]

Newton used these Queries to talk around the subject of optical phenomena, proposing that if light consists of tiny particles, then perhaps the different colors separated out when sunlight passes through a glass prism correlate to different sizes of particles? He suggested that the smallest might appear violet, and the largest red. Despite not offering experimental proof, such was his authority that scientists generally accepted Newton's corpuscles. And they were reluctant to relinquish them until the early 1800s, which was when the first really convincing evidence began to emerge in favor of light being a wave. One experiment in particular appeared to clear up, once and for all, any confusion on the matter. It was one of the most important experiments in the history of modern physics, though it was devised by an amateur physicist who disliked the bother of lab work. It demonstrated that light exhibits that most wave-like of properties: interference.

Thomas Young, born in 1773, was a prodigious polymath who learned to read at two and, by the time he was four, had devoured the Bible from cover to cover—*twice*. Some thirty years later, he made a drawing that would send a chill through the heart of any passing moths.

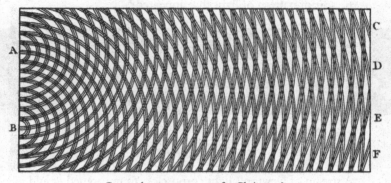

Guess who got a compass for Christmas?

Young presented the illustration as part of a lecture, "Natural Philosophy and the Mechanical Arts," that he gave to the Royal Society in 1807.[6] It was supposed to represent the pattern of wave interaction "obtained by throwing two stones of equal size into a pond at the same instant." Young is said to have gained inspiration for the image from the overlapping patterns of ripples caused by a pair of swans on the pond of Emmanuel College, Cambridge.[7] But he didn't show the drawing just to demonstrate a point about water waves. It was also to illustrate the behavior of light.

Young argued that, as much as ripples water, his drawing represented a beam of sunlight passing through two slits in a piece of card (marked A and B on the diagram), and emerging in the form of waves. As waves, the light ought to spread out from each slit, just like water waves do on passing through a narrow gap in a seawall; this is the familiar wave behavior known as diffraction. Young asserted that if light is a wave, then the overlapping beams emerging from the two slits should interfere with each other as do overlapping water waves, producing regions not of wavier and flatter water but of brighter and darker light. The equivalent of an extra wavy region of water, or constructive interference, would be a bright patch of light; the equivalent of a flatter region, or destructive interference, would be a dark patch. This, explained Young, is exactly what he found when he performed the experiment. When he held a piece of card in the overlapping beams, this is the sort of pattern of light that appeared:*

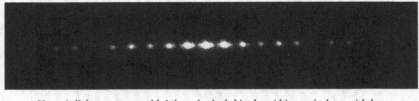

Young's light pattern wouldn't have looked this clear (this one is done with laser light) but the light passing through the two slits certainly interfered to produce lighter and darker regions.

A stream of Newton's light corpuscles, Young argued, could not explain these results, while the bright and dark fringes "may be very easily deduced from the interference of two coincident undulations, which either cooperate or destroy each other." You'd think that he had made a convincing case. How could this be explained if light was made of little particles? Corpuscles plus more corpuscles just makes a lot of corpuscles.

The acceptance of Newton's particles theory was so entrenched that it was more than a decade before Young's argument was taken seriously. Even if other waves, such as those on the water, could be seen to cancel each other out when timed to interfere destructively, the idea that light plus light could equal dark was simply too counterintuitive. Henry Brougham, a young Scottish lawyer and champion of Newton's work, attacked Young's argument viciously in the *Edinburgh Review*, a highly influential periodical that he founded, describing it as lacking "traces of learning, acuteness or ingenuity that might compensate for its evident deficiency in the powers of solid thinking."[8]

....

* In fact, he soon worked out that this effect would only appear if the beams of light from each slit were "coherent"—in other words had matching wavelengths and intensities, and emerged from the slits in time, or "in phase," with each other. This required that the same very bright and pure light source be shone through both slits at once.

The naysayers were only silenced once French engineer Augustin Fresnel formulated the math to support Young's argument. In a presentation to the Paris Academy of Science in 1815, *Light is* he was able to account perfectly for Young's interference *definitely made* fringes with mathematics derived from a wave theory *of waves* of light. Finally, the tide of opinion began to turn, and by the mid-nineteenth century the scientific consensus was that light was most definitely a form of wave.

๏

In December 1900, the German physicist Max Planck inadvertently threw a spanner in the works. He proposed an innocent "what if," which was to become a major headache for anyone arguing *A spanner in* that light is a wave. For five years, Planck had been trying *the works* to come up with a theoretical model for how the light given off by the filament in an electric bulb depended on the temperature of the metal. This was something that the electrical companies were keen to understand in order to make their bulbs more efficient.

For some reason, formulating the relationship between the frequencies of the light and the temperature of the filament proved a considerable challenge. Everyone knows that an iron rod in a blacksmith's forge glows with different colors the hotter it gets: first red, then orange, yellow, and white (after which it generally melts). A range of frequencies is always given off, but the dominant frequency, the one that is brightest, changes with the temperature. As the temperature increases, so does the frequency of the light that is brightest and so the glowing metal changes color. But *how* exactly did the dominant frequency relate to the temperature? No physicist at the time could work out the mathematical formula.

You may be wondering what the big deal was supposed to be. The conundrum might have been of great interest to lightbulb manufacturers, but it is not as if Victorian society was holding its breath in anticipation of an answer to the burning question. But the mathematical solution Max Planck proposed just happened to inspire a twenty-one-year-old Albert Einstein to revolutionize—once again—the world's understanding of light. In fact, through the work of Einstein and others, Planck's innocent mathematical conjecture would eventually reshape our entire picture of the world at an atomic scale.

At this point you have to park the idea that visible light, along with the other electromagnetic frequencies, is a wave in any straightforward sense.

Planck found that by assuming the heat and light emitted by a hot metal takes the form of tiny, indivisible chunks of energy, which he called "quanta," he could accurately predict the frequencies emitted at different temperatures. He devised his system of energy quanta as no more than a mathematical trick to make his calculations fit the experimental data; his conjecture was that the higher the frequency of the light emitted, the more energy was contained in each of these notional quanta. Like other physicists of his era, he believed that light was a form of wave, and that it was only a matter of time before the light and heat emissions from glowing metals could be described in terms of waves.

But a few years later, during his immensely productive *annus mirabilis* of 1905, Einstein proposed that Planck's system of quanta might be more than just mathematical artifice. This then unknown physicist, making ends meet as a clerk in the patent office in Berne, Switzerland, published a paper hypothesizing that electromagnetic *Einstein* radiation actually *is* composed of quantum packets of *wades in* energy.[9] What if, he proposed, these indivisible chunks of energy were real, physical features of light—indeed, of all electromagnetic waves? What if metals, heated enough to emit light, really do give off discrete packets of energy? Were this the case, then perhaps the opposite would hold true: perhaps metals would also *absorb* light in the form of separate chunks of energy? If this could be demonstrated experimentally, our whole understanding of light would be turned on its head once again.

Published in March 1905, this was the first of five groundbreaking papers that Einstein was to write that year. Collectively, their importance can hardly be overstated, for they would form the foundation for the future course of modern physics. Among the papers was the first articulation of the young physicist's theory of relativity but, of the five, only his proposal that light consists of indivisible chunks, or quanta, of energy did Einstein himself describe as truly "revolutionary."[10]

～e～

Einstein hypothesized that if light *were* emitted and absorbed in the form of quanta, then perhaps the different frequencies or wavelengths of light, and which appeared as different colors to us, differed by the amount of energy their quanta contained. The practical way that he suggested this might be demonstrated was by means of a phenomenon, known as the "photoelectric effect," that occurs when certain metals absorb light.

A characteristic of metals is that their electrons are very mobile. This is why they conduct electricity so well. But how mobile they are varies from metal to metal; some are better conductors than others. This electron mobility means that when light falls on a metal, it can sometimes have the effect of "knocking out" *The photoelectric* electrons, causing them actually to fly off the surface. This *effect* is the photoelectric effect. How much they are ejected like this can be measured over time, since by losing the negatively charged electrons, a piece of metal picks up a positive charge. The effect has to be taken into consideration in the design of spacecraft—sunlight falling on its metal fuselage can cause a positive charge to build up, which can affect the instruments. The photoelectric effect is also the principle behind the light sensors at the heart of camera light meters, and the sensors that turn street lights and baby nightlights on when it gets dark. (In these detecting devices, electrons aren't actually ejected from a metal as the light is absorbed, but remain within a "semiconductor," becoming excited from a static state, bonded closely to the atoms, to a fluid state, when the bonds are similar to those in metals and the electrons can flow as a current.)

Einstein's revolutionary paper of 1905 proposed that if light does consist of quanta, then perhaps an electron is kicked out from the surface of a metal by this photoelectric effect when a quantum of light is absorbed. If this were the case, he predicted, the *number* of electrons flying off a metal surface every second would depend on the number of quanta arriving every second, or the *intensity* of the light; while the maximum *speed* with which they flew off would depend on the *energy* in each quantum dislodging them, or the frequency of the light—in other words, its color.

Ten years later, in 1916, Einstein's predictions were confirmed.[11] When red light, which is a relatively low frequency, is shone onto certain metals, electrons would fail to be kicked out from the surface no matter how bright the light was. Green light, on the other hand, which has a middle frequency, readily knocked out the electrons. But they flew off with the same maximum speed no matter how bright the green light was. And the speed of the ejected electrons was greater when even the very faintest glimmer of violet light, which has a high frequency, was shone onto the metal.

This couldn't be explained in terms of light as a wave, but made complete sense if light consisted of these packets of energy, and the amount contained in each quantum depended on the frequency. The photoelectric effect seemed to corroborate Einstein's proposal that light is composed of discrete quanta, rather than spread-out waves. But, just as their equivalents had been so resistant to dropping Newton's conception of light corpuscles in the face of Young's compelling arguments for light being waves, physicists in the first half of the twentieth century were extremely reluctant to relinquish this in place of Einstein's theory that it should, once again, be described in terms of particles. His big idea was universally rejected by his contemporaries, with claims that his "reckless" hypothesis "flies in the face of thoroughly established facts" and that it "is not able to throw light on the nature of radiation."[12]

Einstein, however, was quietly confident. The existence of "the light quanta is practically certain," he wrote in a private letter to *It's particles* a friend in 1916, soon after his predictions about the *again. This is* photoelectric effect had been demonstrated experimen-*getting ridiculous* tally. But it wasn't until 1921 that he received the Nobel Prize in Physics for his 1905 work on the quantum nature of light. Five years later, the light quanta that Planck had proposed (yet had never actually believed in) and Einstein had proven to exist, became known as "photons."

It was all changed again: light *was* made of particles after all.

But what about all that business with Thomas Young's slits? Hadn't his demonstration of the interference of light emerging from a pair of slits shown conclusively that light behaves like a wave? Surely his experiment has proved that it *had* to be a wave. Two particles—whether you call them corpuscles, quanta, photons or subatomic gobstoppers—can't just add up to no particles in the way that out-of-phase waves can cancel each other out.

And what would happen if you started sending these gobstoppers through Young's slits one at a time? It's not as though each one is going to pass through both slits and interfere with itself, is it?

Well, unlikely as it may sound, it actually *is* possible to replicate the results of Young's double-slit experiment—by using filters to reduce the intensity of the light so much that it is sent through the slits photon by photon. Instead of hitting a wall once they've passed through, each photon is detected by a very high-sensitivity camera and recorded as a white dot on a screen.

Initially, the positioning of each arriving electron seems random, but as the white dots build up, something rather weird emerges:

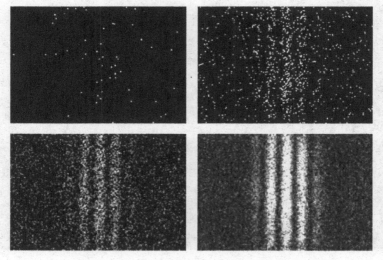

Can you see what it is yet? Photon by photon, a pattern appears.[13]

Light and dark patches that exactly match Young's interference fringes appear, with the majority of the photons hitting the camera where you'd expect bright regions, and hardly any of them where you'd expect dark regions. So, the pattern is the same as you find with interfering waves. It is as though, to quote Paul Dirac, one of the founders of quantum physics, "each photon then interferes only with itself."[14] You may think that what a photon gets up to in the privacy of a darkened box is its own business and no one else's. But Dirac is essentially acknowledging that we haven't a clue why individual photons sometimes behave as waves.

An image assembled from dots like this is reminiscent of pointillism. You might compare it to Paul Signac's 1888 painting of the gently rippling waves in the harbor at Portrieux, in Brittany. Pointillism must be a most tedious method of applying color; the artist builds the scene in tiny dots of paint, monotonous brush point after monotonous brush point. (The most famous painting of the genre, Georges Seurat's *A Sunday Afternoon on the Island of La Grande Jatte*, famously took him two years to complete.) The artist may be positioning by the dots—but who is arranging the photons?

Paul Signac's *The Harbor at Portrieux* (1888).

What hidden hand governs their placement on the screen, seeming at first to be random, but building up over time into a pointillist pattern of wave interference?

It is as though each photon's path is informed by a wave—as if the light behaves like a wave in transit and like a particle when it makes contact with the camera.* "Once the particle has appeared," said the physicist George Paget Thomson, "the wave disappears like a dream when the sleeper wakes."[15] This is the *Who knows? Who cares?* seemingly contradictory world of quantum mechanics, by which the split-personality behavior of electromagnetic waves can now be described mathematically. The theory adequately explains the apparently contradictory behavior of light, but has it brought us any closer to understanding what light actually is? Not according to Richard Feynman, one of the world's greatest quantum physicists: "You are not going to be able to understand . . . That is because *I* don't understand. Nobody does."[16]

Most scientists who have worked in the field of quantum physics have been at pains to say that they find the nature of light as mysterious as everyone else. In 1951, Einstein himself wrote:

All the fifty years of conscious brooding have brought me no closer to the answer to the question, "What are light quanta?" Of course today every rascal thinks he knows the answer, but he is deluding himself.[17]

The split personality of light means it can be described in terms of frequency (red being at the low frequency end of the visible spectrum, blue/violet at the high end) or in terms of the energy of its photons (red-light photons having less energy, blue/violet ones more).

But it isn't just visible light that exhibits this wave-particle duality. All electromagnetic waves do, and can be described in terms of frequency, wavelength or the energy of their photons. The ones with frequencies lower than those of visible light are all

---

* In fact, light is said nowadays to behave as a "quantum-mechanical wave" as it travels and as a particle when it is detected.

those waves so essential to the world of telecommunications: radio waves, microwaves and infrared waves. The ones with frequencies higher than those of visible light are known as ultraviolet waves, X-rays and gamma rays.

Ultraviolet waves

Ultraviolet light waves have wavelengths of between 400nm and 10nm. We call them light waves even though we can't see them, since some of them are visible to other animals. The longer-wavelength ultraviolet waves are emitted by the sun, though not so plentifully as visible light, and are the waves that tan us. But the shorter their wavelengths, the more damaging to our skin are these ultraviolet waves (I could equally have said "the higher their frequencies" or "the more energetic their photons"). They are described as "ionizing" radiation, since they knock the electrons off atoms. When they do this in our skin, they can damage the DNA molecules and turn cells cancerous. Thankfully, our ozone layer absorbs most of the shorter-wavelength, higher-energy ultraviolet waves; but astronauts, outside the protection of the atmosphere, rely on a very thin layer of gold over the surface of their visors to shield them from these waves.

X-rays

X-rays are electromagnetic waves with wavelengths of between 10nm and 0.01nm. Comparatively speaking, the sun doesn't give off many of them; they are, however, emitted profusely by the extremely hot gas clouds that spread out between clusters of colliding galaxies billions of light years across the universe. Rather closer to home, they can be given off by a piece of metal that is bombarded with high-speed electrons. This is how X-rays are produced in order to photograph the fractured bones of your arm. Their higher-energy photons pass easily through your flesh, but not so easily through your bones, and so leave a shadow on a photographic plate on the far side. X-ray photons are also ionizing, and cause cancer even more readily than ultraviolet waves. This is why prolonged exposure to X-rays is best avoided.

And, finally, there are the highest-frequency electromagnetic waves: gamma rays. At less than 0.01nm, these have the shortest wavelengths and the most energetic photons. Out in space, they are produced by celestial bodies far hotter than our sun—such as exploding suns, known as supernovas. Here on earth, they are produced by radioactive materials, which should give you an idea of how dangerous gamma rays can be. But the damage they inflict on living organisms is not all bad news—they are used to great effect in the food industry to eradicate bacteria. And, *Gamma rays* ironically, these deadly photons, so harmful to living cells, save lives. Gamma rays are employed in radiation treatment to kill off cancerous cells in order to stop them self-replicating.*

꿈

In 1924, a French aristocrat showed that the peculiar intertwined nature of waves and subatomic particles extended beyond the world of electromagnetic waves. The thirty-two-year-old Louis, 7th Duc de Broglie (pronounced *broy*), defended his PhD thesis to a committee at the Faculty of Sciences at the Sorbonne in Paris. Its central premise was so outlandish that the examiners were at a loss whether or not to award him his doctorate. De Broglie argued that since Einstein had shown how light waves could be described as streams of tiny particles, now known as photons, perhaps the opposite would hold true. Perhaps streams of tiny particles of matter, such as electrons, or even atoms, each of which has mass (albeit minuscule), could be described as waves.

While the mathematics of de Broglie's work looked sound enough, his conclusion must have seemed ridiculous. With some trepidation, the committee awarded de Broglie his doctorate. One of the examining academics showed it to Albert Einstein, who was impressed: "It may look crazy," he wrote to a fellow physicist, "but it is really sound."[18]

De Broglie's "matter waves" may have seemed absurd, but they were soon shown to exist. In 1927, only three years after his thesis was accepted, two physicists at the Bell Laboratories in New York (which funded basic physics research as well making telephones)

---

* Increasingly, the distinction between X-rays and gamma rays is not based on any arbitrary threshold of wavelength or frequency, but on the way the electromagnetic waves/photons are created. X-rays are produced by the changing energy states of electrons around an atomic nucleus, while gamma rays are emitted by the nucleus itself. (No, that doesn't mean anything to me either.)

found that, just like a beam of light, a beam of electrons would make a diffraction pattern of constructive and destructive interference: when they shot a stream of electrons at a crystal of nickel, they were scattered into concentrated bands. The lattice structure of the nickel acted like the slits in Young's light experiment, though much, much, *much* closer together (somewhat less than half a millionth of the spacing between the slits).

The wave behavior of electrons had been demonstrated. Moreover, by measuring the distance between the bands of the diffraction pattern and taking into account the dimensions of the nickel lattice through which they had passed, the physicists were able to calculate the "wavelength" of this beam of electrons. Their calculation tallied almost exactly with the wavelength that de Broglie had calculated a beam of electrons at that speed should correspond to.

On accepting the 1929 Nobel Prize in Physics for this, de Broglie said, "The electron can no longer be conceived as a single, small granule of electricity; it must be associated with a wave and this wave is no myth; its wavelength can be measured and its interferences predicted."

～e～

You'd be forgiven for scratching your head at all this wave-particle stuff. Does it all really matter? What has the quantum duality of light and subatomic particles ever done for you or me?

Perhaps the most practical technology to have derived from it is the electron microscope—the very device that gave us a glimpse of the mini-Christmas trees on the *Morpho* butterfly wings. Normal optical microscopes could never form an image of the tiny iridescence-producing branches. This has nothing to do with the power of the lenses or the sensitivity of the equipment. It is due to a fundamental limitation of any normal microscope using visible light: it can never have a resolution smaller than half the wavelength of the light itself. So, it is impossible to form an image of cuticle branches less than 100nm thick using visible light with wavelengths of between 400 and 750nm.

The electron microscope

Electron microscopes, on the other hand, achieve a far higher resolution. This can be as little as 0.05nm in the best modern equipment,[19] which is less than the size of an atom. The way these microscopes work depends entirely on the wave-like behavior of streams of electrons. The matter waves can be made to scatter off and diffract around objects as much as light waves do, and can be focused with special lenses to form an image. But with wavelengths of around a millionth that of visible light,[20] the matter waves have the edge when you want to photograph, say, the body hair of a fruit fly.

The resolution of the electron microscope is so amazing that it easily reveals a fruit fly's desperate need for a leg wax.

There are two main types of electron microscope. Scanning electron microscopes measure the electrons bouncing off, or knocked out of, a specimen as it is bombarded with electrons that have been focused into a fine beam. Transmission electron microscopes measure the pattern produced as a broader beam of electrons passes through an extremely thin slice of a specimen. Whether it is the particles passing through or coming off, the whole thing needs to take place in a vacuum, since the electrons tend to be scattered by air molecules. The stream of electrons is drawn off a very hot tungsten filament using a powerful electric field, and then accelerated, often to a velocity approaching the speed of light.

One of the main differences between electron microscopes and optical ones is the type of lenses used. The glass ones in optical microscopes would be completely opaque to a beam of electrons. So instead, very strong magnetic fields serve as the lenses that focus the beam of electrons to form an image.

Another major difference between the two is the effect that a bombarding stream of electrons can have on a specimen compared with a stream of photons. For this reason, a lot of care needs to go into preparing a specimen before it can be imaged. For scanning electron microscopes, this involves coating it in an ultra-thin layer *Goldfinger, on a* of gold so that the electrical charge that builds up from *microscopic scale* the bombardment of negatively charged electrons can be conducted away before it distorts the image. For transmission electron microscopes, it might involve slicing the specimen with a "diamond knife" so that the electrons are able to pass through it—requiring a thickness of less than 0.0001mm (which must need a steady hand).

The waves associated with moving streams of particles are the reason electron microscopes can form images of such incredible magnification. Without knowing about matter waves, we never would have been able to get so intimate with a fruit fly, let alone understand the structural colors of a butterfly.

But de Broglie's discovery has had much more profound consequences than simply letting us see very, very small things in great detail. The fact that he was able to extend Einstein's wave-particle description of light to apply to matter as well is, when you think about it, a very big deal. What he showed mathematically, and has since been corroborated experimentally, is that all tiny particles, when accelerated fast enough, behave as waves—not just electrons, but atoms and molecules, too.

Did you get that? If broken into small enough pieces and sent zooming along fast enough, *all matter has a wavelength*. So, you could say (and, dammit, I think I will) that *everything* is a wave. Perhaps we wave watchers are onto something.

The Ninth Wave

WHICH COMES CRASHING ASHORE

It was now January, and time for that "research trip" to Hawaii. Descending through the fair-weather Cumulus clouds over Honolulu Airport, on the island of Oahu, I felt relieved to be turning my attention back to those most visible, tangible and familiar waves: the ones we find at the seashore.

By the end of the previous year I'd been left with a sense that waves are absolutely everywhere around us and yet, paradoxically, are largely indiscernible to us. We've evolved by paying attention to the information they carry, rather than to the waves themselves. The sea surface provides our most vivid experience of waves; to most of us, ocean waves are the very essence of waviness. As the plane touched down in Honolulu, I found myself remembering those waves on the Cornish shore that had first aroused my interest. Here were different waters, of a different ocean on a different side of the world, and yet I felt an acute sense of returning.

Arriving in brilliant sunshine at Waimea Bay, on the island's North Shore, I clambered down onto the glistening lava rocks of the point. The thunderous roar of the waves before me was overwhelmingly physical. Elvis films, surfing videos and *Hawaii Five-0* hadn't prepared me for this sensory onslaught. Along with the salty tang of sea spray on my lips, the visceral sense of the waves' power *Waves: the* was like a tonic, rousing me from my jet lag. A steady wind *real deal* swept over the palms and the dazzling sand in the bay to my left and blew out across the northwest Pacific waters, from where the heaving, turquoise mountains of water rolled in. These rose into peaks and exploded on the rocks before me, shooting fountains of white water 20ft in the air, which cascaded onto the boulders of the point with a foaming hiss. I took off my shoes in order to feel the pounding force of the Polynesian swell reverberate through the rocks. Here in Hawaii there would be no wave-particle confusions.

~

The following day at Waimea Bay I sat wave watching with Andrew Marr, a big-wave surfer from South Africa. We were taking in the bay from a garden overlooking the "line-up"—a scattering of surfers out in the water, sitting up on their boards and waiting for the next set of waves to come in. The winds I'd felt the day before were still blowing, for these were the warm, moist trade winds that drifted over the Hawaiian islands from the east throughout the year. They rustled the palm fronds above our heads, sending flashes of sunlight dancing across the table in front of us.

Marr was explaining how to surf the enormous swells for which this spot is famous, telling me about the importance of judging the timing and position of the "takeoff," the moment the surfer stops paddling to pick up speed and jumps up onto the board to start riding down the wave face. "The energy suddenly starts to gel at a certain place," he said. "That's where you want to be—waiting at that very point where the energy focuses. If you're in just the right place at just the right moment, just as the swell hits, you've got the perfect entry."

I was nodding enthusiastically, but this was not to suggest any eagerness to grab a board, paddle out there and have a go myself. I would not be experiencing "the perfect entry." I had never stood on a surfboard–hadn't even surfed the piddly waves on the coast near our home in Somerset–so I was not about to try my luck on the heaving monsters that arrive on the shores of Polynesia. My involvement here was deter-minedly that of a detached observer. I knew where I reclined on the Venn diagram of wave watchers and wave riders:

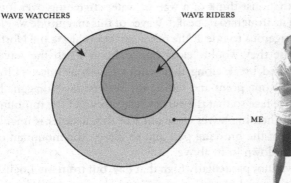

My status in Hawaii was quite clear.

Marr, by contrast, had been surfing since he was four. His father had been a surfer, too, and since his first visit to Hawaii in his early twenties he had been back for a few months each winter.

The bay stretching out before us was one of the most famous surf spots, or "breaks," as they are known, along the North Shore. Indeed, Waimea Bay is the place that brought the world's attention to "big-wave surfing"–which means on waves whose faces are more than 18ft from crest to trough.

From our viewpoint, it had all the attributes of a typical tropical shoreline–white sands fringed by palms and aquamarine water–but the secret to Waimea's fame as a big-wave surf spot was hidden from our view beneath the waves: this was the jagged lava reef that extended below the water from the north point of the bay. When a large swell approaches from the northwest, the part of the wave that rolls into the bay begins to break at its right-hand end (when viewed from the beach). There, in the shallower water where the lava reef extends beyond the point, the wave suddenly slows, steepens and breaks, while the rest powers on through the *Not for the* deep water in the middle of the bay. This means that the *fainthearted* tumbling white water starts on the right and progresses left along the length of the wave. The difference in depth between the reef at the point and the sandy middle of the bay ensures that Waimea can be surfed by anyone brave or foolhardy enough to do so when the wave faces grow as tall as 50ft. If you can't imagine a wave like that, just think of a wall of water stretching over four stories, from its trough to its peak.* Waves of this magnitude would be far too dangerous to ride at the other surf spots along the North Shore, because they would "close out." This is when the waves tumble over and break along their entire length at once, rather than starting at one point and letting the surfers travel diagonally across the wave face so that they always stay ahead of the tumbling white water. When big waves close out like this, a surfer is unable to stay ahead of the breaking part and so escape the mountain of water crashing down from above.

The surf was not particularly high that day, but from my English perspective it seemed blooming enormous. How on earth did the surfers manage to stay upright when shooting down these rolling

....

* In Hawaii surfers have traditionally used a measurement of wave heights that is slightly at variance from elsewhere. In most parts of the world, a wave's height is given in terms of its face height at any moment—in other words, the drop from the crest to the trough in front. The "Hawaii scale," on the other hand, refers to the height of the waves not as they are breaking at the shore, but when the swell is still out at sea and approaching the islands. Since it is away from the shore, the measure applies equally to all the surf spots along the North Shore (while the face heights as the waves break can vary greatly from one beach to the next, depending on the particular depths and seabed features). Typically, the Hawaii-scale swell height is about half the face height as the waves break near the shore. Since it is a bit of a muddle—and makes the surfers seem like a very self-effacing lot since they are always underselling the size of the waves they are riding—I'll stick to the more straightforward international measure of the face height as the wave starts to break.

cliffs of water? According to Marr, this was not the most difficult part. The challenge was judging where to position yourself to catch the wave just as it peaked: "But when you read it right, you paddle out there and, lo and behold, you're in the perfect place to catch it."

Sometimes, if a surfer finds the right position and perfectly times the "pop-up" (which means jumping onto your feet), there's no need even to paddle in order to pick up speed. Standing at the very apex of the wave, the combined weight of surfer and board is enough to send both shooting down the surging precipice of water.

In order to catch the peak like this, the surfers want to position themselves above the shallowest part of the reef. This is where the waves are slowed most dramatically, rising up as they are concertinaed. By lining up landmarks on the shore, Marr explained, they can see if they are drifting off the best spot as they wait for larger waves to roll in.

Watching the surfers out in the bay, I realized that they do a lot more waiting around in the water than I expected. But now it made sense, for the swells travel across the ocean in trains, with groups of larger waves separated by smaller ones. So the surfers paddle and chat through the smaller ones as they wait for a "set" of larger ones to arrive. "Here on Waimea, I use the boils to judge when a good set is coming," Marr explained intriguingly. *Boils before big waves* A boil is a circle on the water's surface, perhaps 13ft across, where the water forms a vortex as it rushes over a peak in the lava reef below, smoothing out the rippled texture of the surface. Surfers try not to ride over the boils, for the rotating water can pull the board in unpredictable ways. Their behavior can, however, be a subtle precursor to the arrival of the big waves.

"You can't even see the set arriving," Marr said, his eyes gleaming with excitement, "but the waves are starting to feel the bottom already. As you sit on your board, you can sense a potential building. It starts with a boil over here and then there's one over there, and then you see a big boil appearing further out."

We gazed out at the scattering of surfers in the line-up, waiting patiently and, presumably, checking their boils. And then, as a set approached from out at sea, the surfers came to life. Each seemed to decide which particular waves to ride at the very last moment, frantically paddling and kicking as the surge of rolling water was about to reach them. A couple timed it just right and powered down the precipice of water, foam crashing at their heels. The whole thing looked petrifying. All physical waves are the transport of energy from one place to another. This terrifying, indomitable energy of the ocean was clearly something these big-wave surfers knew more intimately than most.

⌒

Added together, the coastlines of the four main islands in the Hawaiian chain—Big Island (also known just as Hawaii), Oahu, Maui, and Kauai—amount to 820 miles. Their position right in the middle of the Pacific Ocean, and the fact that they are volcanic islands, which rise from the deep waters without any surrounding continental shelf, means that much of the year they experience "waves of consequence" (surfer-speak for big 'uns). In summer, these are usually around 3ft, but in the winter they're more like 7ft to 10ft, and at times well over 30ft.

The consistency of the dramatic and sizeable swells reaching the shores helps explain why it was on these very islands that the earliest account was recorded of the strange pastime of riding waves on pieces of wood. Lieutenant James King was an officer under the explorer Captain James Cook when two British ships, HMSs *Discovery* and *Resolution*, landed on the islands in 1779. When the crew came ashore in Kealakekua Bay on Big Island, they were greeted as deities. But they clearly outstayed their welcome. A month later, Captain Cook was killed in an altercation with the islanders over a stolen dinghy.

After Cook's death, Lieutenant King assumed responsibility for completing the captain's journal, in which he described taking to the beach when the surf was up:

> . . . *a diversion the most common is upon the Water, where there is a very great Sea, and surf breaking on the Shore. The Men sometimes 20 or 30 go without the Swell of the Surf, & lay themselves*

*flat upon an oval piece of plank about their Size and breadth, they
keep their legs close on top of it, & their Arms are us'd to guide the
plank, they wait the time of the greatest Swell that sets on Shore,
& altogether push forward with their Arms to keep on its top, it
sends them in with a most astonishing Velocity, & the great art is to
guide the plank so as always to keep it in a proper direction on the
top of the Swell, & as it alters its direction.*

Visitors to the islands after Captain Cook's ill-fated voyage
reported that surfing was such an integral part of island life that
villages would empty when a good swell rolled in. Young and old,
male and female, all raced to the beach to play in the surf and
ride the crashing breakers. But the Europeans who came to   *Unwelcome*
the islands through the first half of the nineteenth century   *visitors*
had a disastrous effect on island life. They brought diseases,
which ravaged the population since the islanders had built up
no immunities, and the Calvinist missionaries among them were
less than enthusiastic about the surfing. In fact, they considered it
ungodly—not least because it was performed naked—and discour-
aged islanders from doing it. By 1892, Nathaniel Emerson, a local
physician, wrote that it was "hard to find a surfboard outside our
museums and private collections."[1]

In the early twentieth century, however, the sport experienced a
renaissance that was initially centered at Waikiki, on the southwestern
coast of Oahu, which receives modest swells during the summer. It
wasn't until the 1950s that a small band of intrepid surfers began
to tackle the far more powerful waves that hit the island's North
Shore in the winter. The consistently strong swells that break over
the reefs there made it *the* destination for the emerging world of
professional surfing, and host to dozens of competitions. One of
these, the world's premier big-wave competition, has been held at
Waimea since 1986, and is called "The Quiksilver in Memory of
Eddie Aikau." The Eddie, as it is known—named after the first lifeguard at Waimea Bay, who found fame as a big-wave surfer during
the 1960s and 1970s and died, aged thirty-one, attempting to sail
a canoe from Hawaii to Tahiti—doesn't take place every year. It
is held only in those years when the swells are forecast to produce

Standing-wave surfing: like skateboarding down a high-speed
treadmill that's going in reverse.

wave faces of more than about 50ft. Twenty-eight of the best
big-wave surfers in the world are invited to take part. Wherever
they are, they remain on call through the months of December
and January, ready to jump on a plane to Hawaii when a big swell
is forecast. If the waves aren't mountainous enough, the competition is simply held over until the following year.

After chatting to Andrew Marr, I strolled down to the beach
where some young surfers were digging a trench in the sand that was
blocking the Waimea River from discharging into the bay. Known
as a "berm," this huge mound of sand is pushed up the beach
and into the mouth of the river by the force of the winter waves,
creating a natural dam. During the rainy season, from November
to March, the rainwater flowing downriver from the hills builds up
behind the berm until it finally breaks through and comes gushing
down the beach and into the sea. This breach of the natural dam
happens three or four times a month, apparently, and sometimes
the locals help the water along by digging a channel. Curious, I sat
and watched.

As the river water began to flow, it widened the channel by
gouging the sand away from its sides. Soon it had swelled to a
torrent. Surging down the beach to join the salt water of the surf

zone, the water was kicked upward by the contours of the sandy "riverbed" to form a series of standing waves. These were what the surfers were looking for.

Just like the urban surfers of Munich that we encountered earlier, the youngsters took turns to launch themselves from the banks and ride the current on their stubby foam "boogie boards." It was fascinating to watch them, since I'd only ever seen this done on video.

They zipped over the water, propelled down the face of the standing wave by gravity, as the water rushed up it. It was like someone riding a skateboard on a downward-facing treadmill, the wheels spinning as they remained in the same place. The surfers also rode the waves without going anywhere, carving from side to side across the channel.

As I watched them having so much fun on the flow, I remembered Heraclitus's maxim about not being able to step into the same river twice because there is always new water flowing *A cure for* down it. Heraclitus was rather a miserable fellow who'd *misanthropy* despised his fellow citizens, claiming that they would "do well to hang themselves, every grown man of them." Might he have been happier, I wondered, if he'd been a surfer? Less time going on about flux and a bit of time riding it might have done that particular pre-Socratic grump a world of good.

Sadly for Heraclitus, surfing had yet to be invented.

⌒

Sunset Beach Fire Station lies a mile northeast along the shore from Waimea Bay, across the Kamehameha Highway from the Foodland supermarket, where North Shore surfers buy their cans of Red Bull. In one of the station's parking bays was an impressive yellow and chrome fire truck, with a monster-sized surfboard secured to its top, and branded in a matching yellow livery. When the lifeguards clock off at the end of the day, the Honolulu Fire Department takes over. For this was a rescue board; at 10ft, it dwarfed the 6.5ft boards I'd seen most surfers carrying around.

"When I learned to surf, there were no short boards," Captain Jim Mensching told me. "It took two of us kids to carry a board down to the beach." Mensching, now fifty-two, had been a fireman on the North Shore for thirty-two years. His family moved to Hawaii when he was seven and he's been surfing since he was twelve.

"If one of my friends had use of a board, four or five of us might take turns to ride it," remembered Mensching. He further explained that this was in the 1970s, before the invention of the leash; tied around the surfer's ankle, this stops the board from going AWOL in a "wipeout." "To fetch the board, we'd have to bodysurf a wave into the shore."

Although I'd heard of bodysurfing, I wasn't sure what it involved. The answer was very little. "All you need to bodysurf is a pair of flippers, and you don't necessarily even need them," explained Mensching. "When Captain Cook arrived in the Hawaiian Islands, he found more people body-surfing than using boards. It's the purest form of wave riding— how seals and dolphins do it. They're the best bodysurfers, without a doubt."

At its simplest, bodysurfing means putting your arms to

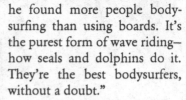

Who needs a pole to slide down when you've got a board to ride?

your sides and planing along the face of the wave. People often stick their arms out ahead of them so that when they get pushed "over the falls" they don't break their necks on the bottom. *Don't break your neck over the falls* Going over the falls is surf-speak for when a surfer is unfortunate enough to fall from the face of a wave as it pitches forward and comes crashing down in an explosion from above; it is the most dangerous way to wipe out, especially on big waves, since the force of such a huge mass of falling water risks the surfer being dashed onto the reef or shore below.

"The idea is to ride along the steep face of the wave at a diagonal, rather than straight ahead," Mensching continued. "You want to plane along from the peak of the wave to the shoulder and, ideally, get covered as it throws and breaks over you." As with bodysurfing, so with boardsurfing. The most sought-after waves are those that start to break at one peak, where the crest of water arcs forward and over toward the trough, and this breaking part travels down the length of the wave. Surfers stay just ahead of the breaking portion, always at the steepest part of the smooth face, traveling diagonally along the front of the wave as it propels them forward. They can even allow the tube to catch up with them so that they disappear just inside its mouth, the canopy of water being thrown over their heads so that they are "in the green room" as they zip along.

The challenge of bodysurfing is to do this without a board. In effect, your body becomes the board. In Hawaii, Mensching explained, it can be very difficult to generate enough speed to stay ahead of the waves, since they are generally so powerful and fast. Remember, the surfer is not being swept along in a current, but is traveling over the surface of the water with the wave, always trying to stay on the slope. And a surfer's body causes a lot more drag as it passes through the water than does the base of a surfboard floating along the top. Bodysurfers will sometimes hold one hand out in front of them to act as a sort of hydrofoil. By planing on an outstretched palm like this, they can lift their torsos out of the water, reducing resistance and creating more speed.

They stretch their other arm behind their back. Mensching demonstrated this, adopting a pose like that of a fencer. "You want to keep this arm out of the water, so you reach behind and sort of hold it above the wave. You can switch hands, do spins and maneuvers like that. Some bodysurfers will do one spin after another."

Once the use of leashes became commonplace, no one needed to ride waves in to shore to catch their boards; so were many people still surfing without boards like this? "Well, when my son and I go bodysurfing, I'm sometimes dismayed to see many of the other bodysurfers are gray-haired old fellas like me."

Besides the leash, the other main reason for the decline in bodysurfing was the introduction, in the 1970s, of the short, foam boogie boards. These are now the most popular board for beginner-level surfing, being easier to ride than regular surfboards. The immediate thrill of zipping along on them has diverted interest away from board-free surfing. "My son took to it just because the boogie boarders were getting all the waves," explained Mensching. "It's frustrating to try and bodysurf at breaks where everyone is on a boogie board. You finally get a wave coming right to you and someone paddling back from just having caught one will turn and drop in on you." "Dropping in" is the cardinal sin. It means catching a wave in front of someone already riding it, thereby forcing them to bail out.

Mensching learned to bodysurf before he used a board, riding waves at the age of seven on Sandy Beach Park, over on the south side of the island. What was it like, I asked him, to be immersed in the energy of the wave? Surveying the sparkling Pacific beyond the fire station, he replied: "It's sort of like flying through the water."

～

Enough talk. I decided it was time to have a go on a surfboard myself, so I called upon "Sunset Suzy," who teaches beginners like me down the coast at Haleiwa. This is a spot where the surf is gentle even in the winter months. Haleiwa means "home of the iwa bird," a large seabird also known as the Great Frigatebird, which is almost always on the wing, only alighting to sleep or to attend to its nest. Unusually for a seabird, the frigate lacks much of the waterproofing oil needed to keep its feathers dry, so it rarely lands on the water. Instead, it plucks fish from the surface, intercepts flying fish in midair or harries other seabirds until they drop their own catches. Because it spends so much time in the air, the bird has learned how to gain lift with minimal effort when flying over the ocean: it makes use of the waves of air as the wind blows over the undulating water. Skimming just above the sea surface, it flits in and out of the troughs between the water waves, rising to take advantage of the lifting air when it needs to. With longer wings, relative to its weight, than any other bird, the frigate is extremely fast and graceful on the waves.

I'll spare you the details of my surfing lessons with Sunset Suzy. Let's just say that I am not a natural. In fact, I am an un-natural. Did I whine earlier about being deaf in one ear? I didn't have an eardrum in my right ear from the age of six months until *Humiliation,* I had an operation to insert one when I was twenty-one. *plus excuses* To avoid infection, I had to prevent water getting in that ear, so when I did swim I only ever did the breaststroke. I blame my pathetic surfing on the absence of water time during the formative years, and on not having great balance (also derived from my bum ear). As well as that, of course, I lack the iwa bird's economical limb-to-weight ratio.

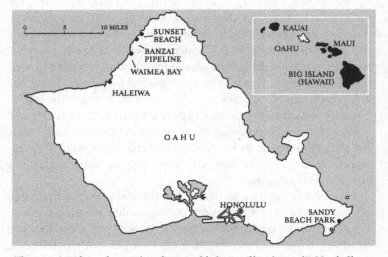

These are just four of more than forty world-class surf breaks on the North Shore.

⌒

After my humiliation on the surfboard, I thought I'd check out bodysurfing. I drove to Sandy Beach Park—Sandy's, as it is known—where Captain Jim had learned to bodysurf. Sandy's lies in the shadow of the steep, striated slopes of the imposing Koko Crater. This is a "tuff cone," a precipitous ring of basalt slopes that surround a hollow interior, formed as huge quantities of ash rained down from a volcanic vent hole during the island's formation 3 million years ago.

Sandy's is just along from the cove where Burt Lancaster rolled amid the rushing waves with Deborah Kerr in *From Here to Eternity*. Since the 1950s, it has been one of the most popular bodysurfing spots in the world. Barack Obama, who grew up in Honolulu and frequented Sandy's as a teenager, was photographed bodysurfing there as a presidential candidate in 2008.

Stepping from the car and onto the scalding sand at the beach's fringe, I could see immediately that the waves were of a totally different character from those at Waimea. Being on the southeastern corner of Oahu, it was in the island's shadow with regard to the winter swells arriving from the north. The waves were not only noticeably smaller, but they also seemed to break in a different way. Rather than starting to break at one end and peeling down the length of the wave as at Waimea, the waves at Sandy's seemed to curl over more abruptly, a greater length of wave crest pitching forward in one go. This is because the seabed here differs from that at Waimea. Sandy's is known as a "shorebreak" because the waves' behavior is dictated by the steep sandy bottom, rather than by any underwater reef. And though these waves were clearly smaller than the ones I'd watched on the North Shore, the steep gradient of the seabed seemed to make them slap down near the water's edge in a rather violent manner.

No one was out on a surfboard, but there were plenty of people on boogie boards. Others were bobbing around wearing flippers. I watched one woman launch herself forward on a breaker, her hand out in front of her, planing along on it just as Mensching had described. But as soon as she lifted the other arm behind her for

From Here to Eternity, with added slaps.

balance, the wave crashed down upon her. She emerged from the frothing white water left behind the wave. The ride had lasted all of three seconds.

This was not exactly a spectator sport. Besides the brevity of the rides, there is the fact that most of the surfer is hidden from view under water. There is none of the drama of the big-wave surfers on Waimea crouching on their brightly colored boards as they speed down the faces of the rolling mountains. For this reason it *Surfing, as* is hard to imagine bodysurfing competitions being shown *Nature intended* on television. Nor is there much scope for bodysurfing fashion. Many of the male surfers just wore Speedos, presumably because they didn't want trendy surf shorts adding to their drag through the water. And, of course, with no boards there's nowhere to put sponsorship logos. Yet I liked the simplicity: the idea of a person, a wave, and nothing more.

The bodysurfer I'd watched emerged from the foaming surf like Ursula Andress in *Dr. No*, and walked up the golden sand toward me, carrying a pair of conch shells. Actually, she looked rather exhausted, and was carrying a pair of flippers. Her name was Shelly O'Brien. She lived on the outskirts of Honolulu, was born on Oahu

What to do when you forget to take your surfboard to the beach.

and raised in a family of surfers. As a girl she spent her weekends bodysurfing. "Now," she told me, "it's all I want to do."

I asked Shelly why Sandy's is so good for bodysurfing. "First, it has a sandy bottom, which is perfect since it's dangerous to body-surf over a lot of coral. Also, the way the waves roll in from deep water and then just shelf up like this is good."

Sandy bottom or not, however, the beach is a notorious accident spot. "The scary thing is that the water looks so inviting. You park your car, and in two minutes, you're in the surf. But it's lethal. This is the number-one beach for spinal injuries." The problem is with swimmers unfamiliar with the style of waves here. "They can pitch you onto three inches of water so that you land on your neck."

I told her that I liked the lack of paraphernalia. "Yeah, it's just your body." She smiled, adding, "I've always said, the bigger the belly, the better the bodysurfer. You watch the old-timers down at *Stop looking at my rudder* Point Panic, or wherever, and the ones with the tummies can do some spectacular things—they're almost like rudders." I found myself locking Shelley in a stare, and realized I was subconsciously hoping to hold her gaze just long enough for the thought to pass before she checked out my own rudder.

c

As I tried to sleep that night, I could hear the waves pounding onto the rocks. Up there, in a small cabin on the peninsula overlooking Waimea Bay, they sounded quite different from out on the beach. The high-frequency hiss of the bubbling white water was missing. All that remained was the low boom of ocean dashing again and again against the headland like distant thunder. I imagined the broad, invisible sound waves diffracting around the house that stood between me and the shore. I'd be in the house's shadow as far as the jittery short-wavelength sounds were concerned, but the long ones would hug the walls and find their way through the night air to my window.

The ocean began to sound like a snoring beast, each microseismic breath sending tiny vibrations up my bed. The last sound to assail my consciousness as I drifted off was the whomph, whomph, whomph of chopper blades. Why on earth would a helicopter be out at that time of night?

c

The next day, I heard that a surfer named Joaquin Velilla had gone missing. In the late afternoon he'd paddled out at a surf break farther along the North Shore and never returned. It had been a Fire Department chopper that I'd heard during the night; they'd searched until 2 a.m., then resumed at daybreak. The Coast Guard had continued through the night using night-vision goggles, but Velilla had yet to be found.

Velilla had paddled out to a break called Banzai Pipeline, the most infamous surf spot in Hawaii. The waves are particularly beautiful: when the swell is large, they pitch forward in dramatic arcs as they pass over the basalt reef, forming large, hollow tubes, or pipes. But for all its beauty, Pipeline is one of the most dangerous breaks in the world. On average, one surfer a year dies there. The waves break in such shallow water that any surfer unlucky enough to go over the falls can be smashed against the reef by the enormous weight of water falling from above. Over the eons, the pounding has smoothed the lava rocks into irregular shapes, full of holes and crevasses. I was horrified to hear that, in a bad wipeout, a surfer can be pushed down into a hole by the force of the wave, never to re-emerge. The wave faces at Pipeline were around 25ft on the evening that Velilla had gone missing.

The perils of Pipeline are partly due to the waves' proximity to the shore. They break in this dramatic fashion just 230ft from the water's edge. This makes it easy for inexperienced surfers to paddle into the thick of the action, but Velilla was no beginner. He knew the break well. A resident of Oahu, he had moved to the island from Puerto Rico five years earlier—a surf immigrant, drawn, like so many Hawaiian residents, by the allure of the waves. The search continued throughout the day but to no avail and, the following morning, Velilla's surfboard was washed up on the beach; its leash had snapped near the ankle harness.

~C~

Apparently the slightest differences in the shape and profile of a surfboard can have a huge effect on how it rides over the waves. The surfers I spoke to seemed to place an inordinate amount of *The dark art* emphasis on the most minor features of their boards—the *of shaping* exact length, whether the base was flat or curved, the position of the tail fins. To be honest, I didn't believe such things could really matter that much, so I made an appointment to meet one of the local surfboard makers, who are known as "shapers."

Jeff Bushman is one of the best shapers on the North Shore, and has been making boards for a living since 1982. Self-taught, he had shaped more than three thousand boards before he even watched anyone else make one. He confessed to having no real idea of how boards work. "It is just an intuitive thing that I've built up over the years," he told me. And, with 30,000 boards under his belt over the past twenty-five years, Bushman has had some time to develop this intuition.

Sometimes the idea for a new design only comes after a nap, as when he was working on a board for local surfer Pancho Sullivan: "I woke up in the middle of the night with a brainwave and thought,

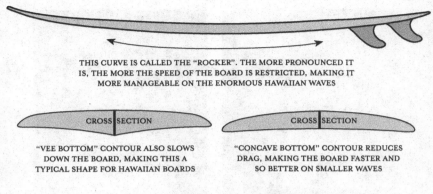

THIS CURVE IS CALLED THE "ROCKER". THE MORE PRONOUNCED IT
IS, THE MORE THE SPEED OF THE BOARD IS RESTRICTED, MAKING IT
MORE MANAGEABLE ON THE ENORMOUS HAWAIIAN WAVES

CROSS SECTION

CROSS SECTION

"VEE BOTTOM" CONTOUR ALSO SLOWS
DOWN THE BOARD, MAKING THIS A
TYPICAL SHAPE FOR HAWAIIAN BOARDS

"CONCAVE BOTTOM" CONTOUR REDUCES
DRAG, MAKING THE BOARD FASTER AND
SO BETTER ON SMALLER WAVES

Surfboards designed for big-wave surfing, such as in Hawaii, are shaped to travel
slower than those for use on smaller waves. This makes them less difficult to
control down the vertiginous wave faces of the North Shore.

that's it. And I got up the next morning knowing what to do." The
board he shaped and then developed in conjunction with Sullivan
was to "revolutionize the boards used at Sunset." But Bushman
didn't mean he'd come up with a new color scheme to paint it, for
the craft of board shaping is not about making them look pretty.
The best shapers are constantly experimenting with dimensions and
profiles to help surfers extract the right amount of energy from the
waves. In fact, shapers like Jeff make boards not only for individual
surfers but for the specific breaks.

"The average surfer here would have a 'quiver' of between eight
and twenty surfboards," he explained, from which they would select
the board best suited to the conditions and location on a particular
day. "These seven miles along the North Shore are unique because
we have so many reefs close to the beaches and each one focuses
the energy of the waves differently. Each one creates different wave
patterns."

Did the boards for Hawaii differ from others? "The surf here
has enormous power," he told me, "so most of the boards we're
building are designed to contain the speed of the wave." This is
done by making the bottom of the board slightly convex when
viewed side-on. In other words, if you stood it on a flat surface,
you could make it rock slightly from head to tail, like a seesaw. This

"Here's one I made earlier": North Shore surfboard shaper, Jeff Bushman, shows me the camber of his bottom profile.

"rocker" makes the board slower over the water, and therefore easier to control on the very steep faces of the huge Hawaiian waves. It also makes the board easier to turn. For big waves, the rocker is often combined with a slight V shape to the bottom, when viewed from an end, the board being thicker down the center, at the apex, and thinner toward either side.

Elsewhere, the surfboard requirements might be the opposite: "If you make a board for a small wave—say in England or the east coast or Japan—you're building a completely different design because you want to *increase* speed." Viewed from their ends, these boards tend to have a concave bottom. The central line of the base is ever so slightly higher than the sides, forming a gentle hollow under the board where air mixes with the water, which gives it lift and reduces the drag as it travels down the wave. "With those boards," Bushman explained, "you need to generate speed because the waves don't have so much power." Were Jeff not such an apparently reasonable fellow, I might have suspected him of looking down on our English waves.

So how did the boards designed for the breaks along the North Shore differ? "Waimea needs a giant board," he replied, without hesitation, "because it takes a massive swell. You might have

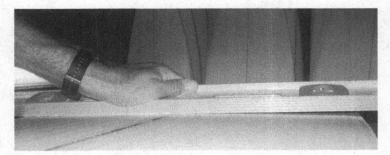

Take a look at the V on that baby.

twenty-foot waves,* which are moving so fast there is no way you can generate enough speed paddling on a small board to get onto them. The Waimea wave will just pass right under you or it will throw you." These boards, known as "guns," are at least 10ft long. Since they spread your weight over a larger area, they are more buoyant and so have less drag.

Sunset, I learned, requires shorter boards—more like 8ft. "The way the reef at sunset sits on the coastline means that it picks up pretty much all the swells that reach the North Shore." For this reason, the break is one of the most consistent spots for waves with 13–20ft faces, which are a little smaller than some of the other breaks along the North Shore. Like many surfboards, the Sunset board tends to have three fins in a triangular formation at the back "to help give it drive, direction and stability." This triumvirate of little rudders serves two purposes. The central, rear fin adds *A board for* stability in maneuvers, since it stops the rear of the board *every break* slipping sideways during a turn. The two side fins, ahead of the rear, might be angled slightly for a Sunset board in order to compress the water as it passes between them, providing lift and giving speed on smaller waves.

And then there is Pipeline . . . "A Pipeline board, for the same rider, would be thinner and at least half an inch narrower; it would also be a little shorter than the Sunset board, with a more pronounced

* Bushman was using Hawaii scale here, meaning the size of the deep-water swell. The equivalent faces, from trough to peak, of the breaking waves would be 30ft.

V contour to the bottom. We've learned that this slower bottom works unbelievably at Pipeline because you're harnessing the speed and you're controlling the energy." This helps the surfers hold back on the enormous barreling waves so that the tube engulfs them, emerging moments before it crashes down on them. "It's the ultimate maneuver—actually being inside the wave, you know, in there feeling that energy. That's kind of a universal feeling in surfing."

～c～

It was time for me to visit Pipeline. When I arrived, flags fluttering from poles in the sand announced that the Rockstar Games Pipeline Pro was in full swing. This is a boogie-boarding competition, which used to be the world championship but is now just one of the several events that comprise the sport's World Tour.

The fans were scattered along the beach, some perfecting a seamless blend of sunbathing and sport watching, others shouting *Flashy moves* encouragement to the surfers. Everyone's eyes were on the fifteen boogie boarders, who were shooting down the faces of the waves, performing 360° spins and doing aerial jumps and flips on their foam boards. The voice on the PA whooped! and yeahed! The waves were pitching over into huge curling tubes. It was thrilling to watch, but I couldn't stop thinking about Joaquin *Somber thoughts* Velilla's disappearance at this very beach two days earlier. What could have happened? Might his body be stuck in some crevice beneath the waves—were the boogie boarders, with their aerials and 360s, inadvertently dancing on his grave?

The swell arriving here had been generated by the winds of a winter storm system that had traveled across the North Pacific from near the east cost of Japan, 3,500 miles to the northwest. These waves had taken two days to reach the North Shore, from where the storm tracked westward, some 500 miles north of the island. I'd never seen such clearly visible wave trains. The breakers arrived at the shore in sets with gaps in between. Perhaps four or five big waves would roll in, each fifteen seconds from the last. This is when the boogie-board champions went crazy, performing aerial spins and "barrel rolls," which involved rushing up the inside face of the tube and flipping right over with the canopy water to land back on their board at the trough. And then for a few minutes the waves would be smaller, and the surfers would regroup and wait. From our slightly elevated perspective on the beach it was easier to see the sets rolling in, so people would whistle frantically to let the surfers know when a group of big ones was on its way.

I decided to stop looking at the boogie boarders' flashy maneuvers and just concentrate on the waves. And this is what I saw.

Out at sea, the swell was a succession of gentle undulations, progressing toward the beach in a sedate and orderly fashion. If you'd only seen these more distant, deeper-water waves, you wouldn't have thought that they'd end up growing as mightily as they did. But this swell was approaching the reef at Pipeline with most of its open-ocean energy intact. The waves hadn't passed through the shallower waters of any continental shelf. They were a steady, relentless train of energy, powering straight for the shore.

There are actually three reefs at Pipeline, each of which affects the progress of the waves. Beyond the one closest to the beach, which is about 230ft from the water's edge, and where the boogie boarders were doing their thing, there is a second, about the same distance again, where the waves break, and which is surfed when the swell is higher. And there is a third reef, at about 920ft, which is known as a "cloudbreak" (when the swell is particularly large, the breaking waves throw up a spray of white water that looks like low clouds). While the swell wasn't large enough for the waves to break at the outer reefs, the wave energy was still focused by them. Where the waves slowed as they passed over the outer reef and lagged behind slightly, the faster parts of the wave on either side curved inward. This was refraction at work—the waves changing direction on account of changing speed.

But all the action was taking place at the point where the waves reached the nearest reef. Here, the gentle arcs of deep water sharpened into crisp, rippling ridges. Slowed dramatically as the water suddenly shallowed, the crests also gathered into a peak at a certain point along the shore. This is where it first began to break. As I watched, I was transfixed by this heaving, trembling mountain of water, poised at the very moment of pure potential.

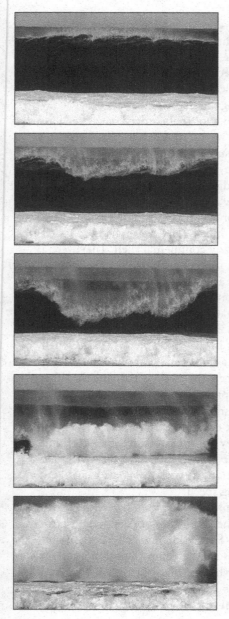

The trade winds, which blow, on average, on one out of every two days through the winter, were blowing from the east, which meant that they were blowing offshore at this part of the coast. Though they cause choppy surface wavelets out at sea, they have the opposite effect on the rising faces of the waves—they smooth out the surface, while also helping to lift the huge ridge of water higher. The winds also lifted a feathering spray from the very tops of the waves, throwing it behind the advancing wave, like a fine mane of mist, to land back on the water in its wake.

Just a split second before breaking, the wave face seemed to change color. The top third suddenly shifted from deep green to an aquamarine. At the same time, the ripples that textured its surface were drawn out, as if combed into long, even ridges down the face, accompanied by beautiful striations that must have been garlands of tiny bubbles.

Now the lip began to tumble. I watched this moment again

The trade winds lifting a delicate mane of spray from the lip of the breakers at Pipeline.

and again as each new wave arrived, never tiring of its effervescent splendor. The white water would start at one point along its length, where the wave rose more than elsewhere, the ridge already lined with white on account of the mane of water lifted by the wind. And there the water began to plunge forward. As the sunlight *A barrel of* gleamed over the lengthening striations down its face, *aquamarine* fingers of white water thrust toward the shore. Behind them rolled a huge barreled roof of aquamarine. This curved over, landing ahead of the wave in an explosion of white foam on the solid reef below, shooting into the air twice as high as the wave itself. The tubular hollow formed by the canopy was open at both ends. As more and more of the wave pitched forward between them and came crashing down in front, the ends peeled away from each down the length of the wave, one to the right and one to the left.

One of the champion boogie boarders now tore into view, but his ride drew my attention to a feature of the breaking waves that I had never noticed before. As he shot down the face, he put a foot into the water to slow his progress just enough to disappear into the mouth of the tube. And there, while inside the hollow of cascading water, he seemed to . . . well . . . explode. A huge puff of water was ejected from the opening, shooting out sideways along the length of the unbroken part of the wave.

There were no gasps of horror from the onlookers on the beach. A second after the explosion of water within the tube, the surfer came shooting out, as the wave collapsed behind him into a mass of tumbling, foaming white water. Now I remembered what this was— I'd been told about it by some of the surfers. It is called the "spit." At a certain point, the air within the tube becomes compressed as the canopy of water comes crashing down and, having nowhere else to go, shoots out of the open end. Surfers look to emerge from the tube just as it spits.

What power and beauty as the waves transformed before us. This, I remembered, was their transition into shock waves. The booming sound and the pounding reverberations, which I could feel through the sand, were the swell's energy being dissipated into the surround-ings. I was just thinking about how the water must be warmed slightly by all the churning turbulence, when, whoa! What was that?

A competitor being spat out of a wave.

Just as a big wave was passing, a tall, slim, silver-haired guy, who'd been bobbing around on the back of a Jet Ski, slid into the water, took a couple of strokes to gather speed, placed his left hand onto the surface in front of him and started bodysurfing. As he traversed the towering face of the wave he held his right arm high behind him. Then, as the wave tumbled over, he drew both arms in beside him and rode the bubbling white water like a seal. He did it with such grace, such ease, almost nonchalance—as if the idea of catching the wave had occurred to him at the very last moment.

Reaching the shore, when there was no more to be had from the swell, he moved over to join the rip current. This is where the water that is pushed up the beach with the successive waves finds a route to flow back out to sea in a continuous stream. It carried him effortlessly out to the Jet Ski, where he lifted himself back on and continued to watch the championship. He looked like the sort of person for whom the word "dude" had been invented.

I had to speak to this guy, so I hung around like a stalker, waiting to accost him as he carried his flippers up the beach at the end of the last heat. When he told me his name, Mark Cunningham, I recognized it immediately. Captain Jim Mensching had *A legendary* described him as one of the best bodysurfers in the world. *bodysurfer* Shelley O'Brien considered him a legend, "synonymous with bodysurfing," and compared him to a dolphin. By pure chance I had stumbled across the one guy who could tell me better than any other what it was like to ride these hulking waves without the aid of a board of any sort.

The search for Joaquin Velilla had been called off after four days. His body was never found. When I called Cunningham to arrange a meeting, he told me he was going to the memorial service the following day, which was being held on the beach at Pipeline. He suggested we meet there.

⏤

When I arrived at the service, the palms lining the beach were bathed in the deepening orange of the afternoon sun. Joaquin Velilla's friends were unwrapping dishes of food and setting out drinks. Crews from the local TV news stations were setting up their tripods as the surfers milled about, many wearing traditional Hawaiian leis around their necks, strung with vibrant pink and

Excuse me? Can I have a word?

purple plumeria flowers. Others hung the fragrant necklaces over the noses of their boards.

In spite of Pipeline's fearsome reputation, and the fact that more surfers have died here than at any other surf spot, Cunningham told me, when I met up with him, how much he disliked the hype: "The media love to play up the dangers of Pipeline," he said. "The surf movies are always, you know, 'the razor-sharp coral at Pipeline is just waiting to slice the surfer up who's had a bad wipeout,' but it's not really like that at all. It is just a hard-breaking wave that pounds down on a shallow, smooth reef with lots of gullies and trenches in it."

Well, I thought to myself, perhaps I'll have a go myself. Not.

Now fifty-one, Mark had recently retired. For almost twenty years he'd worked as a lifeguard here on Ehukai Beach Park, based in a lifeguard tower overlooking the world's most famous wave. "Between doing rescues and first aid and trying to prevent accidents," he said, "I was, literally, a professional wave watcher." In fact, it occurred to me that he must have spent more time watching the waves at Pipeline than any other living person.

Cunningham showed little interest in board surfing. "Occasionally I'll stand up on a board," he told me, "but I'm a one-trick pony, really, when it comes to the ocean." I asked him to describe what exactly he was doing with his body when he body-surfed. He paused: "I've tried to explain it for years and years, and I'm afraid I just can't. It really upsets me that I can't give a good description of it."

I loved the fact that this bodysurfing god was unable to express his actions in words. Of all the people I had asked about body-surfing, Cunningham was the least pretentious: "You have to be in harmony with the waves. It's sort of like dancing with them, really. A friend called it 'downhill swimming.' I like that description. To most people, bodysurfing is jumping up at the right moment and being pushed along by the white water on a friendly beach somewhere. What a handful of us do here is a more advanced sort of bodysurfing. We're trying to ride not the white water, but the face of the wave. It's still just riding a wave, but we're doing it with slightly bigger ones."

Lost for words

Slightly bigger ones? Mark Cunningham was clearly a master of understatement.

~c~

Joaquin Velilla's fiancée, Mariela Acosta, arrived with his family and his closest friend, who made a short but moving speech about Velilla's life. Mariela added that Joaquin had been a man of few words: were he there with us now, he'd just be standing smiling at everyone.

The close band of surfers who knew and loved him were going to honor their friend with a traditional Hawaiian "waterman ceremony." "Please remember to retain the string from your leis," Velilla's friend called as everyone moved down to the beach. "We don't want them left in the water."

After a prayer at the water's edge, the surfers climbed onto their boards and began to paddle out. Velilla had worked as a surfboard shaper, and some now rode boards that he had made. *A waterman* About a hundred yards out, they arranged themselves into *ceremony* an enormous circle, sitting upright on their boards and facing inward. I counted more than forty-five surfers, all of them holding hands.

Velilla's fiancée and his best friend both paddled out to join them. The surfers floated in silence for a few moments, looking down in contemplation at the water lapping around them. Then they shouted his name and slapped cupped hands down by their boards, hurling jets of water high into the air, like a crown of fountains that rocked with the heave of the northwesterly swell.

I was reminded of a Hawaiian poem called "*Na Nalu*," or "The Waves," that I had read the day before:

> *And from the beginning of life to its end . . .*
> *The ocean mirrors the passages of existence on earth.*
> *And even death is but a dark wave that carries the body away in its*
>   *swift current . . .*
> *Out to calm waters where it is dissolved*
> *So preparation can begin for re-birth in a new form.*[2]

Watching all this, I asked Cunningham if he had ever wished for more recognition from the general public for something he loved doing. "No. It's a sort of perverse appeal of the sport for me that there will never be a Bodysurfing World Tour or bodysurfing media and magazines." Would he not have liked a lucrative sponsorship deal? "Perhaps I could have pursued that a little more, but it's just really not in my nature. I don't think it's in the nature of the sport."

We stood next to each other and looked out to sea. The circle of surfers had broken up. Some had paddled back to the shore. A few had joined the line-up, presumably deciding that they would ride the waves in Velilla's honor. The waves continued to roll in, plunging with such graceful arcs I could almost have forgotten their terrifying power.

"Do you want to have a go at bodysurfing?" Cunningham asked, gesturing at the ocean in front of us. Beside us were two signs. One read: "Warning: Strong Current—You could be swept away from shore and could drown," the other: "No Swimming."

"OK," I replied.

We walked down to the foaming ocean fringe. I carried the pair of flippers Cunningham had lent me. He was tanned, and, at fifty-one, had the body of a thirty-year-old. In my late thirties, I had the body of, say, a fifty-one-year-old.

"Where we are right now," he had told me before, "is the impact zone of the world. It is where the land and the sky and the sea meet and exchange energy." Quite a bit of that energy was about to be exchanged through me, I suspected. Still, I felt a reckless exhilara-

What was I thinking? tion as we entered the water, and I reassured myself with the thought that, were I to get into trouble on Pipeline, Mark Cunningham would probably be the one person I would want to have next to me.

"We'll go in the water over there"—he pointed—"ride the current out and head round in front of the sandbar, at the shoulder of the waves."

I detected a degree of disappointment—or was it concern?—when he saw that I was not a strong swimmer. Surfers paddling

Me and Mark Cunningham, the bodysurfing legend.

past yelled greetings to Cunningham. I felt completely out of my element, out of my depth.

While we were swimming out in the rip current, Cunningham explained that I had to duck right under when a wave approached in order to avoid being churned up in the powerful turbulence as it broke. They came, one after another, like a head-on collision of double-decker buses, each with a driver asleep at the wheel. "Never turn your back to the waves when you are in the water," Mark shouted. "That's the number-one rule in Hawaii."

Cunningham showed me the buildings on the shore I should stay level with to avoid drifting over the shallow reef itself where the waves broke strongest. They had lulled slightly—we were between sets—and as we waited I asked what sort of wave I should try to "have a go" with.

"This is a good one," he said, "in fact, I think I might just—" I had to duck under the advancing crest and, when I resurfaced, he had gone.

Where was he? I trod water. I took a breath and dived under another huge breaker. I was swept up its vertical face, which was the height of a small house. Coming up for air behind the breaker, I felt a shower of spray land on my head. The graceful mane of spray that I'd admired from the shore, lifted from the crest by the wind, now seemed more like a mocking gesture—the wave equivalent of a bucket of water pouring down from the top of a door.

And there was Cunningham. He'd ridden the wave all the way in. I realized that I hadn't been able to see him because he'd been surfing at the wave's front while I, left behind, had been looking at its back.

When he'd caught the rip tide back out to where I was, Cunningham tried to help me pick up enough speed to match that of the wave. I'd feel myself lifted by the gathering water, as it surged up from behind. As I frantically kicked my legs, I would for a moment be staring down what seemed an enormous precipice of water. But then it would roll on below me. Only a couple of times did a wave propel me just long enough for me to feel its immense energy.

They were too fast for me to catch, and I felt foolish, frustrated and drained. This was ridiculous. Who did I think I was kidding? I'm no bodysurfer, I thought to myself, spitting out yet another mouthful of salt water.

And then I gave up worrying. I stopped thinking of the waves as wild beasts to be harnessed or mounted. It occurred to me that all the fear was preventing me from actually feeling the waves. I was so focused on avoiding ending up pummeled onto the reef that I had become numbed to the waves themselves.

My heart was pounding as I waited for the next breaker. I thought of the tiny, spiraling waves of electrical impulses that spread with each beat through the muscle in my chest; of the sound of the *I go with* crashing breakers spreading as invisible, crisscrossing pres-*the flow* sure waves through the breezy coastal air; of the sunlight falling on my head, the warming waves of shifting electromagnetic potential scattering on the ocean surface around me. The unstoppable draw of the heaving water made me feel that the next wave was going to be larger than the others. My surroundings, and I with them, began to shift under the spell of the arriving energy, now reaching the end of its long journey across the Pacific. Just as the vertical wall of water began to lift me, I ducked deep below the surface. This time, I just wanted to feel at one with the water,

Wayne Levin's *Mark Under Breaking Wave* shows Mark Cunningham
bodysurfing from below.

to allow myself to be the medium possessed by the energy of the
swell. If I couldn't ride the monster, I would momentarily be a part
of it. The wave broke directly above me.

As the crest barreled, the maelstrom of white water looked
glorious. For a split second I saw a sky of storm clouds. These
were clouds of tiny bubbles churned into the water by the crashing
turbulence above. No longer fighting the wave, I just floated there
beneath the surface, looking upward, my body pulled back and
forth by the rotating sweep of the water. I felt, from the inside,
the death throes of this particular Pacific swell, and watched the
sunlight filter down through the dancing, effervescent surface of
the wave.

NOTES

WAVE WATCHING FOR BEGINNERS

1 King James Bible, Genesis 1: 1–2.
2 Ibid., 7.
3 Coleridge, S. T., "The Rime of the Ancient Mariner," *Lyrical Ballads* (1798).
4 Swinburne, Algernon Charles, "Laus Veneris" (1866).
5 Conrad, J., *The Nigger of the "Narcissus": A Tale of the Sea* (1897).
6 Plutarch, *Morals: Natural Questions*.
7 Bede, *Historia ecclesiastica gentis Anglorum* (*Ecclesiastical History of the English People*), Book III, 15 (AD 731).
8 Franklin, Benjamin, "Oil on Water," A letter to William Brownrigg, November 7, 1773, *Philosophical Transactions* 64: 445 (1774). The letter is also available online at www.historycarper.com/resources/twobf3/letter12 .htm.
9 Munk, W. H., and Snodgrass, F. E., "Measurements of southern swell at Guadalupe Island," *Deep-Sea Research* 4, no. 4 (1957).
10 Snodgrass, F. E., et al., "Propagation of Ocean Swell across the Pacific," *Philosophical Transactions of the Royal Society A*, Mathematical, Physical & Engineering Sciences (1966).
11 "One Man's Noise: Stories of an Adventuresome Oceanographer," written, produced, and directed by Irwin Rosten, University of California Television (1994). Available at www.youtube.com/watch?gl=GB&v=je3QvqNdHl0.
12 Ruskin, John, "Of Waters as Painted by Turner" (1843), *Modern Painters*, Book 2, Chapter 3.
13 Emerson, Ralph Waldo, "Seashore," first published in *May-Day and Other Pieces* (Boston: Ticknor & Fields, 1867).
14 Sophocles, Antigone, trans. R. C. Robb, *The Complete Greek Drama*, vol. 1, ed. W. J. Oates and E. G. O'Neill (New York: Random House, 1938).
15 Whitman, Walt, "That Long Scan of Waves," from the cycle "Fancies at Navesink," in *Leaves of Grass* (1891–92), first published in the magazine *Nineteenth Century*, August 1885.
16 Hogarth, William, *The Analysis of Beauty*, Chapter VII, "Of Lines" (1753).
17 Ibid., Chapter XVII, "Of Action," and Chapter V, "Of Intricacy" (1753).
18 Arnold, Matthew, "Dover Beach," *New Poems* (London: Macmillan, 1867). Apologies for having removed the line breaks.

THE FIRST WAVE

1 Miller, David J., "Heart," in *The Oxford Companion to the Body*, ed. Colin Blakemore and Sheila Jennett (Oxford: Oxford University Press, 2001).
2 Wu, J. Y., Huang, Xiaoying, and Zhang, Chuan, "Propagating waves of activity in the neocortex: what they are, what they do," *Neuroscientist* 14 (5): 487–502 (October 2008).
3 Berger, Hans, "Über das Elektrenkephalogramm des Menschen," *European Archives of Psychiatry and Clinical Neuroscience* 87, no. 1 (1929).
4 Sterman, M. B., and Friar, L., "Suppression of seizures in an epileptic following sensorimotor EEG feedback training," *Electroencephalogr Clin Neurophysiol* 33: 89–95 (1972).
5 Sterman, M. B., MacDonald, L. R., and Stone, R. K., "Biofeedback training of

the sensorimotor electro-encephalogram rhythm in man: effects on epilepsy," *Epilepsia* 15: 395–416 (1974).

6   Sterman, M. B., and MacDonald, L. R., "Effects of central cortical EEG feedback training on incidence of poorly controlled seizures," *Epilepsia* 19: 207–22 (1978).

7   Quoted in Robbins, Jim, "A Symphony in the Brain: The Evolution of a New Brain Waves," *Biofeedback* (New York: Grove Press, 2000).

8   Lubar, J. F., and Bahler, W. W., "Behavioral management of epileptic seizures following EEG biofeedback training of the sensorimotor rhythm," *Biofeedback Self Regul* 7: 77–104 (1976).

9   Lubar, J. F., et al., "EEG operant conditioning in intractable epileptics," *Arch Neurol* 38 (11): 700–704 (1981).

10 Lantz, D., and Sterman, M. B., "Neuropsychological assessment of subjects with uncontrolled epilepsy: effects of EEG biofeedback training," *Epilepsia* 29: 163–71 (1988).

11 Sterman, M. B., "Basic concepts and clinical findings in the treatment of seizure disorders with EEG operant conditioning," *Clin Electroencephalogr* 32 (1): 45–55 (2000).

12 Sterman, M. B., and Egner, T., "Foundation and practice of neurofeedback for the treatment of epilepsy," *Applied Psychophysiology and Biofeedback* 31, no. 1 (March 2006).

13 Arns, M., et al., "Efficacy of neurofeedback treatment in ADHD: The effects on inattention, impulsivity and hyperactivity: a meta-analysis," *Clin EEG and Neuroscience* 40 (3): 180–89 (July 2009).

14 Egner, T., and Gruzelier, J. H., "Ecological validity of neurofeedback: modulation of slow wave EEG enhances musical performance," *Neuroreport* 14: 1221–24 (2003).

15 Professor Farquharson's film, and that of a camera shop owner who was based nearby, have become classic viewing for civil engineering students on the dangers of failing to take the effects of wind into account when designing structures. The film of the bridge twisting like an elastic band is unforgettable, and is easily found on the likes of YouTube.

16 Details of the 1940 Tacoma Narrows Bridge disaster, including eyewitness reports, can be found on the website of the Washington State Department of Transport: www.wsdot.wa.gov/tnbhistory.

17 Hood, Thomas, "The Death-Bed" (1831). *The Poetical Works of Thomas Hood* (London: Frederick Warne & Co., 1890).

THE SECOND WAVE

1   Holmes, Oliver Wendell, "The Philosopher to His Love" (1924–25).

2   Smith, Stevie, "Not Waving but Drowning" (1953).

3   Hrncir, Michael, et al., "Thoracic vibrations in stingless bees (*Melipona seminigra*): resonances of the thorax influence vibrations associated with flight but not those associated with sound production," *Journal of Experimental Biology* 211: 678–85 (2008).

4   This figure is given in Moravcsik, Michael, *Musical Sound: An Introduction to the Physics of Music* (New York: Paragon House, 1987).

5   Vitruvius, *The Ten Books on Architecture*, trans. Morris Hicky Morgan (London: Humphrey Milford, Oxford University Press, 1914), Chapter III: "The Theatre: Its Site, Foundations and Acoustics," Section 6.

6   Ovid, *Metamorphoses*, trans. Anthony S. Kline (Borders Classics, 2004), Book III.

7   Ibid.

8 "Harassed Rancher Who Located 'Saucer' Sorry He Told About It," *Roswell Daily Record*, July 9, 1947.
9 "The Roswell Report: Fact vs. Fiction in the New Mexico Desert." Headquarters of the U.S. Air Force (1995).
10 Popper, A. N., and Fay, R. R., *Sound Source Localization* (Berlin: Springer, 2005).
11 Heffner, R. S., "Comparative study of sound localization and its anatomical correlates in mammals," *Acta Oto-Laryngologica* 117, issue S532 (1997).
12 Mason, Andrew C., Oshinsky, Michael L., and Hoy, Ron R., "Hyperacute directional hearing in a microscale auditory system," *Nature* 410: 686–90 (April 5, 2001).
13 Payne, Roger S., "Acoustic location of prey by barn owls (Tyto Alba)," *Journal of Experimental Biology* 54: 535–73 (1971).
14 Detailed accounts of the swell-analysis skills of the navigators of *Oceania* are given in Lewis, D., *We the Navigators: The Ancient Art of Landfinding in the Pacific* (Honolulu: University of Hawaii Press, 1994).
15 Aea, H., *The History of Ebon* (1862). The Hawaiian Historical Society 56th Annual Report, 1947.
16 De Brum, R., "Marshallese Navigation," *Micronesian Reporter* 10: 1–10 (1962).

THE THIRD WAVE

1 Feynman, Richard, "Fun to Imagine," BBC Television (1983).
2 Mozart, Leopold, "A Treatise on the Fundamental Principles of Violin Playing" (1756).
3 Carlyle, Thomas, "Essay on Burns" (1828).
4 Brontë, Emily, *Wuthering Heights*, Chapter 1 (1847).
5 Kerouac, Jack, *On the Road* (1957).
6 Wolfe, Tom, *The Electric Kool-Aid Acid Test* (1968).
7 Leary, Timothy, *Flashbacks* (1983).
8 This, at least, is how Aristotle described Pythagoras's view in *On the Heavens*.
9 Devereux, P., and Jahn, R. G., "Preliminary investigations and cognitive considerations of the acoustical resonances of selected archaeological sites," *Antiquity* 70: 665–66 (1996).
10 Devereux, Paul, *Stone Age Soundtracks: The Acoustic Archaeology of Ancient Sites* (London: Vega, 2001).
11 Watson, Aaron, and Keating, David, "Architecture and sound: an acoustic analysis of megalithic monuments in prehistoric Britain," *Antiquity*, vol. 73, no. 280 (1999).
12 *NASA-STD-3000: Man-Systems Integration Standards*, Revision B, July 1995. vol. 1, 5.5.2.3.1.
13 Broner, N., "The Effects of Low Frequency Noise on People—A Review," *Journal of Sound and Vibration*, vol. 58, no. 4 (1978).
14 Sheldrake, Rupert, *A New Science of Life: The Hypothesis of Formative Causation* (London: Blond & Briggs, 1981).
15 Agar, W. E., Drummond, F. H., Tiegs, O. W., and Gunson, M. M., "Fourth (final) report on a test of McDougall's Lamarckian experiment on the training of rats," *Journal of Experimental Biology* 31: 307–21 (1954).
16 Sheldrake, Rupert, *A New Science of Life: The Hypothesis of Formative Causation* (London: Icon Books, 2009), p. 119.
17 Maddox, J., "A Book for Burning?" *Nature* 293 (1981).
18 *Santa Fe New Mexican*, September 20, 2008.
19 See www.o-a.info/mmca/explain4.html.

THE FOURTH WAVE

1 Open University, *Waves, Tides and Shallow-Water Processes* (Oxford: Butterworth-Heinemann, 1999).
2 Frost, Robert, "Sand Dunes," from West-running Brook (1928).
3 The Open University, *Waves, Tides and Shallow-Water Processes* (Oxford: Butterworth-Heinemann, 1999).
4 Emerson, R. W., "Seashore" (1857).
5 "Technical Assistance to the People's Republic of China for Optimizing Initiatives to Combat Desertification in Gansu Province." Asian Development Bank, June 2001. Available at www.adb.org/Documents/TARs/PRC/R90-01.pdf.
6 Ellis, L., "Desertification and Environmental Health Trends in China," a China Environmental Health Project Research Brief (April 2007). Available at www.wilsoncenter.org/topics/docs/desertificationapril2.pdf.
7 Brown, L. R., *Outgrowing the Earth: The Food Security Challenge in an Age of Falling Water Tables and Rising Temperatures* (New York: W. W. Norton & Co., 2005).
8 Sugiyama, Y., et al., "Traffic jams without bottlenecks—experimental evidence for the physical mechanism of the formation of a jam," *New Journal of Physics* 10 (2008).
9 The studies are: Kerner, B. S., "Three-Phase Traffic Theory," and Helbing, D., et al., "Critical Discussion of 'Synchronized Flow,' simulation of Pedestrian Evacuation and Optimization of Production Processes." Both studies in *Traffic and Granular Flow '01*, ed. M. Fukui, Y. Sugiyama, M. Schreckenberg, and D. E. Wolf (Berlin: Springer, 2003). Sugiyama, Y., et al., "Traffic jams without bottlenecks—experimental evidence for the physical mechanism of the formation of a jam," *New Journal of Physics* 10 (2008).
10 See, for example, Treiterer, J., and Myers, J. A., *Transportation and Traffic Theory*, ed. D. J. Buckley (New York: Elsevier, 1974), p. 13.
11 Quotations are from: Hippolytus, *Refutation* (IX.10.4); Porphyry, *Quaestiones Homericae*, on *Iliad* XIV, 200; Plutarch, *Consolatio ad Apollonium*, 10. Translation: Loeb Classical Library edition, 1928.
12 Plutarch, *Quaest. Nat.* ii, p. 912, and Plato, *Crat.* 402 a.
13 Aristotle, *The Physics*, Book 8, Chapter 3 (350 BC).
14 Also according to Aristotle, *Meteor* ii. 2, p. 355 a 9.
15 Russell, Bertrand, *A History of Western Philosophy* (1946).

THE FIFTH WAVE

1 The details of Sergeant Emme's experiences were reported in Okie, Susan, "Traumatic Brain Injury in the War Zone," *New England Journal of Medicine* 352, no. 20 (May 19, 2005). Also, his own account of the events is at www .sermonstore.org/2004/Soldiers/D-Emme.html.
2 Hoge, Charles W., et al., "Mild Traumatic Brain Injury in U.S. Soldiers Returning from Iraq," *New England Journal of Medicine* 358, no. 5 (January 31, 2008).
3 Tanielian, Terri, and Jaycox, Lisa H., eds, *Invisible Wounds of War: Psychological and Cognitive Injuries, Their Consequences, and Services to Assist Recovery* (RAND Corporation, 2008).
4 Mellor, S. G., "The relationship of blast loading to death and injury from explosion," *World Journal of Surgery* 16, no. 5 (September 1992).
5 Moss, W. C., King, M. J., and Blackman, E. G., "Skull Flexure from Blast Waves: A Mechanism for Brain Injury with Implications for Helmet Design," *Phys. Rev. Lett.* 103, issue 10, 108702 (2009).

6 Winchester, Simon, *Krakatoa: The Day the World Exploded* (London: Penguin, 2003).
7 Goriely, A., and McMillen, T., "Shape of a Cracking Whip," *Phys. Rev. Lett.* 88, issue 24 (2002).
8 Shearer, Peter M., *Introduction to Seismology* (Cambridge: Cambridge University Press, 1999).
9 Holmes, A., and Duff, D., *Holmes' Principles of Physical Geology* (London: Routledge, 1993).
10 Ibid.
11 Dutton, C. E., "The Charleston Earthquake of August 31, 1886," U.S. Geological Survey, 9th Annual Report, 1887–1888.
12 Versluis, M., Schmitz, B., von der Heydt, A., and Lohse, D., "How snapping shrimp snap: through cavitating bubbles," *Science* 289: 2114–17 (2000).
13 Lohse, D., Schmitz, B., and Versluis, M., "Snapping shrimp make flashing bubbles," *Nature* 413 (October 4, 2001).
14 Tennyson, Alfred, Lord, "Break, Break, Break" (1834).

THE SIXTH WAVE

1 This is the sum of the viewing figures for all the individual matches according to the FIFA World Cup TV viewing figures: Final Competitions 1986–2006. Available from www.fifa.com.
2 Oldroyd, B. P., and Wongsiri, S., *Asian Honey Bees: Biology, Conservation, and Human Interaction* (Cambridge: Harvard University Press, 2006).
3 The shimmering defense strategy is described in Kastberger, G., Schmelzer, E., and Kranner, I., "Social Waves in Giant Honeybees Repel Hornets," *PLoS ONE* 3 (9): e3141. doi:10.1371/journal.pone.0003141 (2008).
4 Siegert, F., and Weijer, C. J., "Three-dimensional scroll waves organize Dictyostelium slugs," *Proc Natl Acad Sci USA* 89 (14): 6433–6437 (July 15, 1992).
5 Farkas, I., Helbing, D., and Vicsek, T., "Mexican Waves in an Excitable Medium," *Nature* 419: 487–90 (September 12, 2002).
6 Farkas, I., and Vicsek, T., "Initiating a Mexican Wave: An Instantaneous Collective Decision with Both Short- and Long-Range Interactions," *Physica A* 369: 830–40 (2006).
7 Ibid.
8 Barklow, W., "Hippo talk," *Natural History* 5/95: 54 (1995).
9 Heppner, F. H., and Haffner, J., "Communication in Bird Flocks: An Electro-magnetic Model," in *Biological and Clinical Effects of Low-Frequency Magnetic and Electric Fields*, J. G. Llaurado, A. Sances, Jr., and J. H. Battocletti, eds. (Springfield: Charles C. Thomas, 1974).
10 Selous, E., *Thought-transferrence (or what?) in Birds* (London: Constable, 1931).
11 Potts, W. K., "The Chorus-Line Hypothesis of Manoeuvre Coordination in Avian Flocks," *Nature* 309 (May 24, 1984).
12 See www.gohuskies.com.
13 See www.gohuskies.com/trads/020498aad.html.
14 See www.krazygeorge.com.
15 "Making Waves Over the Cheer," *Dallas Morning Herald*, November 15, 1984.
16 www.gameops.com/interview/krazy-george/p=2.
17 Dawkins, R., *The Selfish Gene* (Oxford: Oxford University Press, 1989).
18 Nicholls, H., "Pandemic Influenza: The Inside Story," *PLoS Biol.* 4 (2): e50 (February 2006).

19 Benedictow, O. J., *The Black Death 1346–1353: The Complete History* (Woodbridge: The Boydell Press, 2004).
20 The latest figures can be found at http://ecdc.europa.eu/en/healthtopics/H1N1.
21 Cage, S., "Flu drug Tamiflu boosts Roche sales in Q3," Reuters News Agency, October 15, 2009.
22 "CDC Estimates of 2009 H1N1 Influenza Cases, Hospitalizations and Deaths in the United States, April–December 12, 2009," Centers for Disease Control and Prevention, January 15, 2010: www.cdc.gov/h1n1flu/estimates_2009_h1n1.htm.

THE SEVENTH WAVE

1 http://news.bbc.co.uk/1/hi/england/lancashire/4364586.stm.
2 Bartsch-Winkler, Susan, and Lynch, David K., "Catalogue of Worldwide Tidal Bore Occurrences and Characteristics," U.S. Geological Survey Circular, 1022 (1988).
3 Rowbotham, F. W., *The Severn Bore* (Newton Abbott: David & Charles, 1970).
4 Details of the bores around the world are listed on the website of the Tidal Bore Research Society at www.tidalbore.info. All the bore heights listed here are the ones recorded there.
5 Cartwright, David Edgar, *Tides: A Scientific History* (Cambridge: Cambridge University Press, 1998), p. 18.
6 Koppel, T., *Ebb and Flow: Tides and Life on Our Once and Future Planet* (Toronto: Dundurn, 2007).
7 Lifei, Zheng, "Special Supplement: When the waters engulf the sun and sky," *China Daily*, September 8, 2007.
8 Shi, Su, "Watching the Tidal Bore on Mid-Autumn Festival" (1073). From *Su Dong-po: A New Translation*, trans. Xu Yuan-zhong (Hong Kong: Commercial Press, 1982).
9 Ibid.
10 Arrian of Nicomedia, *The Anabasis of Alexander or, The History of the Wars and Conquests of Alexander the Great*, trans. with commentary from the Greek of Arrian the Nicomedian by E. J. Chinnock (1884), Book 6, Chapter XIX.
11 "Future Marine Energy: Results of the Marine Energy Challenge: Cost competitiveness and growth of wave and tidal stream energy," Carbon Trust, 2006.
12 "Progress through Partnership," Marine Foresight Panel Report, Office of Science and Technology, May 1997, URN 97/639, paragraph 2.8. See www.foresight.gov.uk.
13 Details of the excavation can be found at www.nendrum.utvinternet.com.
14 Greenwood, J., "A Gazetteer of Tidemills in England & Wales, Past and Present," at http://victorian.fortunecity.com/holbein/871.
15 "Tidal Stream Energy: Resource and Technology Summary Report," Carbon Trust, 2005.
16 Davey, Norman, *Studies in Tidal Power* (London: Constable & Co., 1923).
17 "Tidal Power in the UK: Research Report 4—Severn non-barrage options," an evidence-based report by AEA Energy & Environment for the Sustainable Development Commission, October 2007. See www.sd-commission.org.uk.
18 See www.rspb.org.uk/ourwork/casework/details.asp?id=tcm:9-228221.
19 Friends of the Earth Cymru, "The Severn Barrage" report, September 2007.
20 Lathe, R., "Early tides—response to Varga et al.," *Icarus* 80 (2006).
21 Dorminey, B., "Without the Moon, Would There Be Life on Earth?" *Scientific American* (April 21, 2009).

22 Lathe, R., "Fast tidal cycling and the origin of life," *Icarus* 168 (2004).
23 Dorminey, B., "Without the Moon, Would There Be Life on Earth?" *Scientific American* (April 21, 2009).
24 Freedman, R. A., *Universe*, 8th revised ed. (London: W. H. Freeman & Co., 2007), p. 249.

THE EIGHTH WAVE

1 Vukusic, P., et al., "Quantified interference and diffraction in single *Morpho* butterfly scales," *Proc. R. Soc. Lond. B* 266: 1403–11 (1999).
2 Yoshioka, S., and Kinoshita, S., "Effect of Macroscopic Structure in Iridescent Color of the Peacock Feathers," *Forma* 17: 169–81 (2002).
3 Zi, J., Xindi, Y., Li, Y., et al., "Coloration strategies in peacock feathers," *PNAS* 100: 12576–78 (2003).
4 He said it in 1762, according to Boswell, James, *Life of Samuel Johnson*, Part 3.
5 Newton, Sir Isaac, *Opticks* (1704).
6 Young, Thomas, "A Course of Lectures on Natural Philosophy and the Mechanical Arts" (London: J. Johnson, 1807).
7 Bendall, S., Brooke, C., and Collinson, P., *A History of Emmanuel College, Cambridge* (Woodbridge: The Boydell Press, 1999).
8 Brougham, H., "The Bakerian Lecture on the Theory of Light and Colors, by Thomas Young," *Edinburgh Review* I (January 1803).
9 Einstein, Albert, "On a Heuristic Viewpoint Concerning the Production and Transformation of Light," *Annalen der Physik* 17: 132–48 (1905).
10 Letter from Einstein to Conrad Habicht, May 18 or 25, 1905, in Klein, M. J., Kox, A. J., and Schulmann, R., eds, *The Collected Papers of Albert Einstein*, vol. 5, *The Swiss Years: Correspondence, 1902–1914* (Princeton: Princeton University Press, 1993).
11 Millikan, Robert A., "A direct photoelectric determination of Planck's 'h,'" *Phys. Rev.* 7: 355–88 (1916).
12 The first quotation is by Robert A. Millikan in *Phys. Rev.* 7: 355–58 (1916). The second quotation is by Neils Bohr, in his acceptance speech for the Nobel Prize in Physics, 1922.
13 Dimitrova, T. L. and Weis, A., "The wave-particle duality of light: A demonstration experiment," *Am. J. Phys.* 76 (2) (2008).
14 Dirac, Paul, *The Principles of Quantum Mechanics* (Oxford: Oxford University Press, 1958, first published 1930), Chapter 1.
15 Feynman, Richard P., *QED: The Strange Theory of Light and Matter* (Princeton: Princeton University Press (1985), Chapter 1.
16 Einstein, Albert, in a letter to his friend Michele Besso (December 12, 1951).
17 Einstein, Albert, in a letter to Heindrick Lorentz (December 16, 1924).
18 Erni, R., et al., "Atomic-Resolution Imaging with a Sub-50-pm Electron Probe," *Phys. Rev. Lett.* 102: 096101 (2009).
19 McKenzie, D., "The Electron Microscope as an Illustration of the Wave Nature of the Electron," Science Teachers' Workshop 2000, School of Physics, the University of Sydney, Australia.

THE NINTH WAVE

1 Warshaw, M., *The Encyclopedia of Surfing* (New York: Harcourt Books, 2005).
2 This was quoted in Kristin Zambucka, *Princess Kaiulani of Hawaii: The Monarchy's Last Hope* (Green Glass Productions Inc., 1998).

AUTHOR'S ACKNOWLEDGMENTS

I would be lying if I said I'd found this an easy book to write. It was really difficult, and took a lot longer than I'd ever imagined. And I've no idea how I would have managed it at all without the generosity of the following people.

Richard Atkinson, my editor at Bloomsbury, gave huge amounts of his time to help develop the structure of the book and hone my text. He must surely take more care over his books than any other editor working today. Patrick Walsh, my agent at Conville & Walsh, gave up his Christmas holiday to read and vastly improve what was a rather clunky draft. Both to him and Richard, I'm extremely grateful.

There are many others I would like to thank. Chief among them is my wife, Liz. She not only contributed ideas, improvements, and corrections, but also offered support on the many occasions when I felt like giving up. Charles Ullathorne, the physics teacher at Westminster School in London, generously gave the manuscript a read-through to weed out some of the science errors. John Fanning, my father-in-law, kindly read it with an eye for more general errors. Roderick Jackson made pertinent suggestions along the way and, crucially, kept my spirits up. And I am very grateful to the following, who read and commented on individual chapters: Professor Andreas Baas, Jamie Brisick, Guy de Beaujeu, Professor Mark Cramer, Professor Pedro Ferreira, Dr. John Powell, Professor Yuki Sugiyama, and Professor Tamás Vicsek.

I'm also grateful to Natalie Hunt at Bloomsbury for her diligent editing and coordinating work, to Igor Toronyi-Lalic for help with research at an early stage, to Richard Collins for his fine copyediting, to Trish Burgess for her impeccable proofreading, as well as to Jude Drake and Xa Shaw Stewart at Bloomsbury, and Charlotte Isaac at Conville & Walsh.

And then there are those who have made more specific contributions. I'm extremely grateful to Cathy and Peregrine St. Germans for putting me up in Hawaii and introducing me to their friends and contacts there, and also to John Bain, of Waimea Bay, Oahu, for assistance while I was in town. I'm grateful as well to Melissa Foks and Dr. Beverley Steffert for help on brain waves, Tom and Donny Wright for help on the Severn Bore, Veronica Esaulova for help on Mexican wave statistics, and finally, for general assistance, to Alex Bellos, Charles Hazlewood, Josh Hallil, Gulya and Barbara Somlai, and Ron Westmaas.

Oh, and thank *you* for reading it.

All chapter illustrations © David Rooney. Diagrams © Graham White, NB Illustration, as listed on copyright page (p.4). All other diagrams are © Gavin Pretor-Pinney.

10: © Gavin Pretor-Pinney. 12: © Jon Bowles (Cloud Appreciation Society Member 16,267). 22: National Oceanic and Atmospheric Administration. 25: © National Maritime Museum, Greenwich, London. 29: © The Metropolitan Museum of Art/ Art Resource/Scala, Florence. 34: © The Art Archive/Tate Gallery London. 38: © National Maritime Museum, Greenwich, London. 48: © Michael Schrager. 65: © Ed Eliot, The Camera Shop, Tacoma–reproduced with permission. 74: © Kim Taylor/naturepl.com. 91 (both): photographs by US Air Force, from "The Roswell Report" (1995). 92: Reproduced with permission from the *Roswell Daily Record*. 100 (top): © Marco Lillini (Cloud Appreciation Society Member 2,120), marco@ lillini.com, www.lillini.com. 100 (bottom): © Edwin Beckenbach/Getty Images. 101: © The Trustees of the British Museum. All rights reserved. 106: © Gavin Pretor-Pinney. 109: Photo: ESA. 111: © Frank and Myra Fan. 120: © Orpheon Foundation, www.orpheon.org. 127 (both): © Gavin Pretor-Pinney. 132: © Steimer/ ARCO/naturepl.com. 135 (both): © Bruce Odland. 141: Munich Surf Open publicity posters from Grossstadtsurfer 2000 e.V.: www.grossstadtsurfer.de. 145: © Dr Harry Folster (Cloud Appreciation Society Member 20,843). 149: © Gavin Pretor-Pinney. 151: Photograph published in *The World's Work* magazine (Doubleday, Page & Company, 1908). 159: © Professor Y. Sugiyama, the Mathematical Society of Traffic Flow. 169: US Navy photograph by Phan Elliott. 173: © Wiel Koekkoek (Cloud Appreciation Society Member 16,471). 178 (top): US Navy photograph by Ensign John Gay. 178 (bottom): US Navy photograph by Photographer's Mate 3rd Class Jonathan Chandler. 180: © Steve Bly/Getty Images. 183: NASA. 184 (all photographs): NASA. 192: © James Lyle. 193: © Michel Versluis, University of Twente. 200 (all photographs): © 2008 Kastberger et al. 203: © M.J. Grimson & R.L. Blanton, Biological Sciences Electron Microscopy Laboratory, Texas Tech University. 206: © Bob Thomas/Getty Images. 211: @ Richard Barnes, www.richardbarnes.net. 215 and 216: Reproduced with permission from SRO Productions on behalf of George Henderson, aka Krazy George. 224: Photographer: American Red Cross. National Oceanic and Atmospheric Administration/US Department of Commerce. 225 (top): National Oceanic and Atmospheric Administration/US Department of Commerce. 225 (bottom): © David Rydevik–reproduced with permission. 226: © Gavin Pretor-Pinney. 230: Photograph © Richard Kruml, Fine Japanese Prints, Paintings and Books, London. 238: © John Franklin/BIPs/Getty Images. 241 and 242: © Gavin Pretor-Pinney. 245: Photograph by Susan Bartsch-Winkler, US Geological Survey. 264: TEM image © P. Vukusic, University of Exeter. Published in "Quantified interference and diffraction in single *Morpho* butterfly scales", Proceedings of the Royal Society B: Biological Sciences: 1999 July 22; 266 (1427): 1403. 272: © Department of Physics, Columbia University, NY. 277: Reprinted with permission from T.L. Dimitrova and A. Weis, *American Journal of Physics*, Volume 76, Issue 2 (2008). © 2008, American Association of Physics Teachers. 278: Staatsgalerie, Stuttgart, Germany/ Giraudon/ The Bridgeman Art Library. 283: Power and Syred/Science Photo Library. 294: © Gavin Pretor-Pinney. 295: Photo montage, Gavin Pretor-Pinney. 296: © Gavin Pretor-Pinney. 301 and 302: © Greg Rice, www.GregRImagery.com. 306, 307, 310, 312 and 313 (all photos): © Gavin Pretor-Pinney. 319: © Wayne Levin, www.waynelevinimages.com.

Every effort has been made to contact and clear permissions with the relevant copyright holders. In the event of any omissions, please contact the publishers.